MINDING THE CARBON STORE:
Weighing U.S. Forestry Strategies to Slow Global Warming

Mark C. Trexler

WORLD RESOURCES INSTITUTE

January 1991

Kathleen Courrier
Publications Director

Brooks Clapp
Marketing Manager

Hyacinth Billings
Production Manager

Each World Resources Institute Report represents a timely, scientific treatment of a subject of public concern. WRI takes responsibility for choosing the study topics and guaranteeing its authors and researchers freedom of inquiry. It also solicits and responds to the guidance of advisory panels and expert reviewers. Unless otherwise stated, however, all the interpretation and findings set forth in WRI publications are those of the authors.

Contents

Acknowledgments

Many people contributed in significant ways to this report. Lisa Loewen was indispensable in pulling together the many sources of required information and in assisting with the analysis that went into this study. Bill Moomaw guided the analysis from its inception as congressional testimony late in 1988, and Roger Dower assumed the same role upon joining the World Resources Institute in January 1990. Other members of the WRI staff, including Paul Faeth, Bob Repetto, Mohamed El-Ashry, Jim MacKenzie, and Gus Speth, provided valuable reviews and comments.

The analysis profited greatly from reviews by a number of colleagues outside WRI. The author especially appreciates the comments provided by Lynn Wright, Richard Birdsey, Robert Moulton, Pete Emerson, Robert Fischman, Roger Sedjo, Dwight Hair, Mark Harmon, and David Turner. Members of WRI's Policy Panel on Responses to the Greenhouse Effect and Global Climate Change who provided valuable feedback include George Woodwell, Senator Robert Stafford, Margaret Kripke, and Harold Corbett.

The clarity and quality of the manuscript were vastly improved by the editorial efforts of Kathleen Courrier. My thanks also go to my wife Laura Kosloff for her editorial insights. Production of the report was made possible by the efforts of Lisa Loewen, Margot Greenlee, Cindy Barger, Hyacinth Billings, Robbie Nichols, and Allyn Massey. Any errors present despite all this help are solely the responsibility of the author.

M.C.T.

Foreword

After decades of discovery and discussion, scientists are reaching consensus on the ways that human activities are affecting the natural greenhouse effect. In near unison, they estimate that we have already committed the planet to significant warming that may not make itself manifest for some decades.

If we hope to slow—and eventually halt—global warming, the challenge is to control atmospheric concentrations of heat-trapping gases. Carbon dioxide, still the leading culprit, is a particular problem since its concentration is now growing by 0.5 percent per year. All told, human activities give rise to some seven billion tons of carbon in the form of carbon dioxide annually. That looks like a mere blip on the screen when Earth's carbon cycle moves 400 billion tons of carbon around each year, but it is enough to throw the natural cycle out of equilibrium, driving atmospheric carbon content steadily upward—by about 25 percent since preindustrial times. Capping the mercury's resulting climb will require staunching carbon emissions or expanding the so-called sinks that keep carbon out of the atmosphere, or both.

Tree planting, one of many ideas percolating through scientific and policymaking circles about how to store more carbon, is already being tried on a small scale. With WRI's advice, for instance, a U.S. electricity supplier is funding a forestry project in Guatemala to offset the lifetime carbon emissions of its new coal-fired powerplant in Connecticut. President Bush's America the Beautiful proposal aims at planting 10 billion trees over the next decade, and the American Forestry Association's Global Releaf program aims at planting 100 million more trees in cities by 1992. Environmental ministers from 68 countries recently adopted an ambitious goal for the year 2000: to increase global forest cover by 30 million acres a year—a complete turnaround from today's annual *loss* of 30 to 50 million acres. A host of other ideas for storing and displacing carbon are making the rounds, too, from increasing the carbon content of soils, to making durable products out of wood rather than plastic, to substituting biomass for fossil fuels.

Some carbon-storing mechanisms promise many benefits *besides* mitigating climate change. Sustainable forestry and agriculture, for instance, not only protect carbon storage but also maintain such vital resources as soil and water. Such practices are often recommended for the tropics, but they are just as important for the industrial world. The United States and other developed countries could help balance the global carbon cycle if they stepped up existing conservation and reforestation programs and made carbon storage an explicit goal of their forestry and agricultural policies. Doing so might also lend them the moral standing needed to influence tropical countries, which face far tougher decisions in the struggle to meet human needs while controlling greenhouse gas emissions.

Storing more carbon is particularly important for the United States, the world's biggest fossil-fuel user. Many industrialized democracies have already set carbon stabilization or reduction goals. To meet these goals, they will have to cut carbon emissions by using less fossil fuel, but they may also be able to soak up more carbon by expanding carbon "sinks." Although the United States is now resisting the adoption of specific goals and timetables, the pressure to join the international alliance will grow as this country becomes ever more politically isolated among its industrial peers.

For all these reasons, carbon sequestration merits careful examination. In *Minding the Carbon Store: Weighing U.S. Forestry Strategies to Slow Global Warming*, Mark C. Trexler provides economic, political, and technical analyses for each of seven policy options that are feasible for the United States. Major trends and policies must be changed, he argues, and so must the complex incentive structures that currently guide the decisions of private landowners and public land managers. Dr. Trexler maintains that improving U.S. forestry and agricultural policies and practices offers significant potential for storing carbon and displacing fossil-fuel use, though these biotic options are neither as simple nor as inexpensive as commonly believed. Another caveat that emerges from his analysis is that while forestry can help moderate net carbon emissions, increased tree planting cannot compensate for the lack of a U.S. energy policy intended to address the problem of global warming.

Building on testimony submitted before the Senate Agriculture Committee in December 1988, Dr. Trexler assesses the political and economic ramifications of needed changes and the difficulties of bringing them about. He concludes that vigorous pursuit of sound biotic policies can help our leaders forge a transitional energy strategy to combat global warming. The policy recommendations spelled out in this report extend those of such other WRI studies as *A Matter of Degrees: The Potential for Limiting the Greenhouse Effect, Growing Power: Bioenergy for Development and Industry*, and *Energy for a Sustainable World*.

Minding the Carbon Store owes an intellectual debt to the leading experts in international relations, climatology, agriculture, energy, environment, law, trade, industry, and the developing world who serve on WRI's Policy Panel on Responses to the Greenhouse Effect and Global Climate Change. In several meetings over the past two years, the panel deliberated at length over climate change in general and reforestation in particular. Its members offered comments and recommendations crucial to the analysis contained in this report, and their assistance is highly valued.

We would like to thank the Ford Foundation, the Public Welfare Foundation, and the Rockefeller Brothers Fund for their generous support of WRI's research on energy and climate issues. To all these institutions, we owe a debt of gratitude.

James Gustave Speth
President
World Resources Institute

WORLD RESOURCES INSTITUTE
POLICY PANEL ON
RESPONSES TO
THE GREENHOUSE EFFECT AND GLOBAL CLIMATE CHANGE

Mr. Richard E. Ayres
Senior Staff Attorney
Natural Resources Defense Council

Ambassador Richard E. Benedick
Senior Fellow
World Wildlife Fund/The Conservation
Foundation

Mr. Harold Corbett
Senior Vice President
The Monsanto Company

Dr. John Firor
Director, Advanced Study Program
National Center for Atmospheric Research

Dr. Richard Gardner
Professor of Law and International
Organization
Columbia University School of Law

Margaret L. Kripke, MD
Professor & Chairman
Department of Immunology
University of Texas

Mr. C. Payne Lucas
Executive Director
AFRICARE

Dr. Gordon J. MacDonald
Vice President and Chief Scientist
The MITRE Corporation

Dr. Jessica Tuchman Mathews
Vice President
World Resources Institute

Mr. Robert McNamara
Former President, the World Bank

Mr. William G. Miller
President
The American Committee on U.S.-Soviet
Relations

Dr. Michael Oppenheimer
Senior Scientist
Environmental Defense Fund

Dr. George Rathjens
Center for International Studies
Massachusetts Institute of Technology

Mr. Roger Sant
Chairman
Applied Energy Services

Dr. Maxine Savitz
Director, Garrett Ceramic Components Division
Garrett Processing Company

Dr. Joseph J. Sisco
Sisco Associates

Mr. James Gustave Speth
President
World Resources Institute

I. Introduction

The composition of Earth's atmosphere is changing under human influence. Of major concern is the increase in concentrations of the so-called greenhouse gases. At natural levels, these gases keep Earth warm enough to support life, but increasing their concentrations is likely to warm the planet further and take global temperatures outside the relatively narrow band within which today's natural systems and human societies have evolved.

One way to slow and moderate global warming might be to manipulate the planet's biotic carbon cycle. One concept now being discussed is the fertilization of marine ecosystems in the hope that plankton populations would bloom and then sink to the ocean bottom, taking large quantities of carbon with them. Alternatively, deforestation could be slowed (thus slowing the release of carbon to the atmosphere), carbon could be drawn out of the atmosphere by increasing the storage capacity of terrestrial vegetation and soils, and biomass could be substituted for fossil fuels in producing needed energy. More generally, the management of forest, pasture, agricultural, and other lands can be modified around the world to slow the accumulation of carbon dioxide (CO_2) in the atmosphere and at the same time help meet such other policy objectives as sustainable economic development, watershed protection, energy efficiency, soil conservation, and the preservation of biological diversity.

The overall potential and cost-effectiveness of biotic policy options remain poorly understood, partly because what is appropriate in one country might not work as well in another.

The overall potential and cost-effectiveness of biotic policy options remain poorly understood, partly because what is appropriate in one country might not work as well in another. Tropical countries, where ongoing forest loss is most severe, often need to protect and increase forest cover if they want to safeguard watersheds and other natural resources as well as provide energy for their increasing populations. Large areas of marginal or degraded lands that are at least theoretically available for reforestation are also often present in tropical countries. In these same countries, however, competition for available land is often intense, and many still lack the governmental and private infrastructure needed to make large-scale reforestation work over the long run. In contrast, temperate industrialized countries, such as the United States, are not currently facing large-scale deforestation. Similarly, few have large areas of severely degraded or abandoned lands

1

that could support trees. Many temperate countries are better positioned, however, to employ market and management mechanisms to increase carbon storage on lands already being used.

A start in this direction is the president's America the Beautiful program announced in early 1990. Under this program, the Forest Service would coordinate and fund the planting of an additional one billion trees per year on private lands for at least the next 10 years. The program is also intended to endow U.S. cities with greater tree cover. According to various estimates, the program would offset the emission of from 16 to 65 million tons of carbon annually after 10 years of planting—roughly equivalent to one to four percent of current U.S. CO_2 emissions from fossil fuel combustion.

Why do these estimates of the potential carbon benefits of the America the Beautiful program range so widely? And how can forestry-based policy options' effect on global warming be evaluated? To find out, the first step is understanding how trees can help slow global warming and determining whether existing forestry policies are having this effect. The next is reviewing the theoretical potential, the likely costs, and the probable advantages and disadvantages of various forestry-based policy options. Such options include slowing the conversion of forested land to alternative uses, accelerating the reforestation of suitable lands, adapting current forest-management practices to enhance carbon-storage goals, making cities more comfortable and energy efficient through tree planting, substituting biomass-based electricity and liquid fuels for fossil fuel consumption, and increasing the use of wood products.

As the following chapters show, each of these policy options has distinct advantages and disadvantages from a carbon-storage perspective, and no two have exactly the same overall potential for slowing global warming. Indeed, the uncertainties involved in projecting the specific implications and cost-effectiveness of the alternative options are so overwhelming that precise conclusions are out of the question. But one thing does emerge quite clearly from the following analysis: although forestry may be able to partially offset U.S. emissions of CO_2 from fossil fuel combustion, stepped-up tree planting cannot eliminate the need for policies intended to reduce the emissions of carbon dioxide in the first place.

II. The Carbon Cycle and Global Warming

Global warming is believed likely to result from a variety of human activities, with those that release carbon dioxide into the atmosphere topping the list. How fast CO_2 builds up in the atmosphere, however, depends on the workings of the carbon cycle, the largest and best quantified of the various natural cycles that sustain life on the planet. *(See Box 1.)*

The Global Carbon Cycle

When in equilibrium, carbon flows between the various reservoirs of the global carbon cycle roughly balance over the year: carbon emitted to the atmosphere through plant respiration and decay is offset by carbon drawn out through photosynthesis. Over the last 160,000 years (the extent of reliable CO_2 data), the carbon cycle has slowly moved in and out of balance a number of times, with atmospheric CO_2 concentrations ranging between 200 and 280 parts per million (ppm) over the millennia. High concentrations of atmospheric CO_2 characteristically have been associated with particularly warm climatic periods, and low concentrations of atmospheric CO_2 have been associated with cooler periods and the one ice age occurring during this 160,000-year period.

There is little doubt that the carbon cycle is now once again out of balance. In just two centuries, CO_2 concentrations have risen by over 25 percent from approximately 280 ppm in

1750 to more than 350 ppm in 1989. They are currently rising by about 1.4 ppm or 0.5 percent per year on average. Today, the atmosphere contains roughly 160 billion more tons of carbon than it did before industrialization began, reflecting an unprecedented rate of change.

There is little doubt that the carbon cycle is now once again out of balance.

Human activities, including those dating back to well before the industrial revolution, are the primary force behind the carbon cycle's current instability. For centuries before the mid-1700s, for example, CO_2 was released into the atmosphere as people converted temperate forests and grasslands to farmland. Even today, owing mainly to deforestation in the tropics, these biotic releases are probably greater than ever. Estimates of the quantity of carbon released through land-use change in 1980, for example, range from 1.5 to 2.1 billion tons.[1] Even higher numbers are postulated for more recent years as tropical deforestation has accelerated.[2]

Cumulatively, an estimated 150 to 250 billion tons of carbon have been released to the atmosphere since 1860 through land-use change around the world.[3] *(See Figure 1.)* The range of

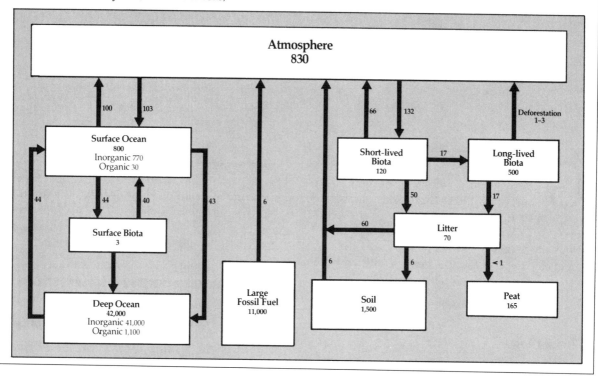

The Global Carbon Cycle   (in Billions of Tons)

Atmosphere
830

Surface Ocean
800
Inorganic 770
Organic 30

Surface Biota
3

Deep Ocean
42,000
Inorganic 41,000
Organic 1,100

Large
Fossil Fuel
11,000

Short-lived
Biota
120

Long-lived
Biota
500

Litter
70

Soil
1,500

Peat
165

Deforestation
1–3

100 103   66 132   17   44   44   40   43   6   50   17   60   6   6   < 1

Human activities, including those dating back to well before the industrial revolution, are the primary force behind the carbon cycle's current instability.

---

estimates is wide because accurately quantifying carbon releases from land-use change is quite difficult. Nobody knows, for instance, exactly how many acres of forest are currently being converted or how much carbon is released per acre from vegetation and soils during conversion. *(See Box 2.)* Complicating

matters further is uncertainty about whether rising CO_2 concentrations are fertilizing remaining forests, particularly in temperate zones, thus offsetting some of the carbon emitted as trees are felled elsewhere in the world.[4]

Although biotic CO_2 emissions could rise to as high as 5.5 billion tons per year early in the next century if deforestation rates increase linearly with population growth,[5] any such increase would be cut short quickly by the virtual disappearance of tropical forests from many countries. But there is nothing to limit growth in the other major source of human-caused CO_2 emissions—namely, the combustion of fossil fuels for energy production. Although cumulative biotic carbon emissions probably exceeded cumulative fossil fuel emissions until just a couple of decades ago, fossil fuel consumption is growing so rapidly that

The global carbon cycle is made up of large carbon flows and reservoirs. Hundreds of billions of tons of carbon as CO_2 are absorbed from or emitted to the atmosphere annually through natural processes. These flows include vegetational photosynthesis, respiration, and decay as well as the oceanic absorption and release of CO_2. But they are dwarfed by the various carbon reservoirs. Empirical measurements show that the atmosphere contains about 825 billion tons of carbon. An additional 800 billion tons are dissolved in the surface layers of the world's oceans. Because they are so heterogeneous, terrestrial carbon reservoirs are much more difficult to measure empirically. Some 1,650 billion tons of carbon are believed to have accumulated in ground litter and soils. Terrestrial organisms, primarily plants, account for an estimated 615 billion additional tons of carbon, though this number is the subject of considerable debate. (See Box 2.) By far, the largest carbon reservoirs are the deep oceans and fossil fuel deposits, which

account for some 42,000 an̲ tons of carbon, respectivel̲

Human activities relea̲ tons of fossil fuel carbon to tn̲ in 1988. Deforestation, mostly in tropi̲ countries, is estimated to have contributed more than two billion additional tons. Only about three billion tons, however, remained permanently in the atmosphere. Until recently, scientists believed that the bulk of the difference was absorbed by the surface layers of the oceans, but they are now increasingly contending that heightened CO_2 levels are "fertilizing" upper-latitude forests, potentially by as much as three billion tons of carbon annually.

Source: World Resources Institute in collaboration with the United Nations Environment Programme and the United Nations Development Programme, 1990. *World Resources Report 1990–91.* Oxford University Press.

the reverse is now true. (See Figure 2.) CO_2 emissions from fossil fuels quadrupled between 1950 and 1980 alone. If nothing is done, says the Intergovernmental Panel on Climate Change, carbon emissions will probably increase by roughly two to three percent annually, going from some 6 billion tons today to more than 8 billion tons by 2000 and to more than 13 billion tons by 2025.[6]

Most countries release more carbon than they take in, but how much and how each contributes to the buildup of CO_2 in the atmosphere varies widely. In many tropical nations, for example, the dominant source of carbon emissions is land-use change. Although this used to be the case in most temperate countries as well, fossil fuel combustion is now the leading source. The United States is no exception to this general trend.

The U.S. Carbon Cycle

Fossil fuel consumption in the United States is well monitored. U.S. carbon emissions from fossil fuel combustion now total some 1.6 billion tons per year, and they are increasing. But although the fossil fuel portion of the U.S. carbon cycle is well quantified, the biotic side of the cycle is much less so. We do know how the 2.2 billion acres of surface area that make up the United States are distributed. (See Figure 3.) Land area and uses have been extensively studied and surveyed, but it is not known whether existing incentives and policies are already having the effect of offsetting some proportion of U.S. fossil fuel emissions by storing additional carbon or whether the reverse is occurring. (See Appendix.) Our inability to predict whether U.S. forest area will continue its 40-year decline illustrates the degree of our

Figure 1. CO_2 Released Through Regional Land Use Changes, 1860–1980

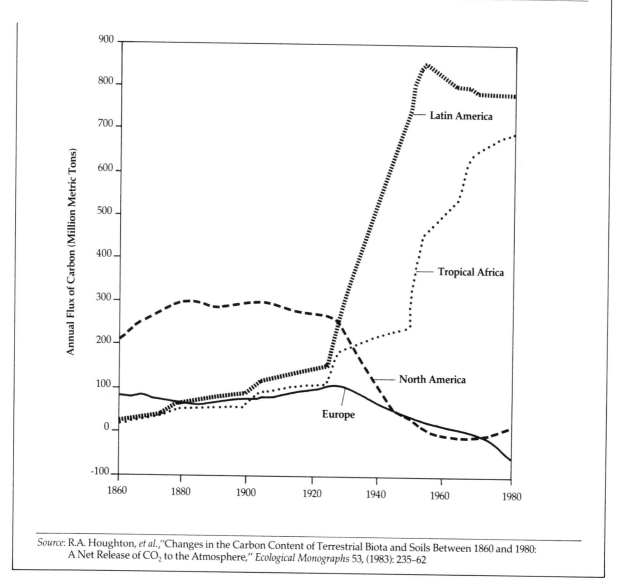

Source: R.A. Houghton, *et al.*,"Changes in the Carbon Content of Terrestrial Biota and Soils Between 1860 and 1980: A Net Release of CO_2 to the Atmosphere," *Ecological Monographs* 53, (1983): 235–62

ignorance. Figure 4 shows trends in U.S. forest area since 1920 and projects alternative scenarios of future forest area. The most optimistic reflects continuation of a short-term trend created by recent implementation of the Conservation Reserve Program, but that program is scheduled to end soon. The intermediate projection reflects the short-term trend without including the Conservation Reserve Program, and the third projection simply extends long-term trends into the future. The United States Forest Service (USFS) does, in fact, project a continuing decline in at least the timberlands component of U.S. forest area.[7] Timberlands are forests that are capable of producing more than 20 cubic feet of merchantable timber per year and that are not "reserved" for uses other than timber production.

Box 2. Estimating Global Biomass Levels

It is commonly assumed that the terrestrial biotic component of the carbon cycle makes up a carbon reservoir of more than 615 billion tons of carbon. Forests no doubt hold most of this total. Recently, however, some scientists have challenged prevailing biomass estimates, arguing that prevailing sampling methods significantly overstate average biomass concentrations, particularly for forests. They note that current estimates of tropical forest biomass, for example, are based on samples totalling less than 75 acres of forest, and that the chosen samples reflect forest ecosystems of above-average biomass density. Using timber volume estimates, which are based on much larger-scale sampling than biomass estimates, Brown and Lugo concluded that the total carbon stored in tropical forests is nearer 250 billion tons, only one-half the approximately 500 billion tons estimated by Whittaker and Likens a decade earlier.

New biome-specific research lends weight to the conclusion that global estimates have been overstated. Recent research on North American boreal forests suggests that the carbon content of above-ground biomass averages from 7 to 10 tons per acre, much less than the 25 to 35 tons-per-acre average in previous estimates. University of California researchers suggest that most older estimates were based on sampling undisturbed mature forest, thus overlooking the lower carbon densities found in disturbed forests. In addition, past estimates used the southern range of the boreal forest, where biomass density is highest.

Significant revisions in global biomass estimates have important ramifications for understanding the carbon cycle and global warming. If the amount of biomass has been overestimated, for example, current deforestation would contribute less to global CO_2 emissions than is often assumed, and the potential role of CO_2 fertilization in moderating growth in atmospheric CO_2 concentrations would probably not be as significant as would otherwise be the case. Scenarios based on lower biomass estimates also suggest that slowing deforestation, and even increasing afforestation, would not have as beneficial an effect on global warming as might be projected under the high-biomass scenario.

Sources: Botkin, D.B., and Simpson, L.G., 1990. ''Biomass of the North American Boreal Forest,'' *Biogeochemistry* 9:161–174; Brown, S., and Lugo, A.E., 1984. ''Biomass of Tropical Forests: A New Estimate Based on Forest Volumes,'' *Science* 223:1290–93; Whittaker, R.H., and Likens, G.E., 1973. ''Carbon in the Biota,'' in Woodwell, G.M., and Pecan, E.V., eds. *Carbon and the Biosphere.* U.S. Atomic Energy Commission. Symposium Series No. 30. National Technical Information Service.

Forests, especially timberlands, are almost certainly the nation's largest storehouses of biotic carbon. Although croplands, pasture, and rangelands cover more area, they store far less carbon than do forests,[8] and urban and developed areas store even less. Domestic forest management practices will therefore significantly influence U.S. carbon stocks and flows in coming years. The role of U.S. forests as carbon reservoirs—if the United States plays a role in international efforts to mitigate climate change—is likely soon to become much more prominent. Some U.S. forests, particularly Pacific Northwest old-growth forest, contain far more carbon per acre than the Brazilian tropical forests whose protection has repeatedly been called for by interest groups in the United States.[9] *(See Box 3.)*

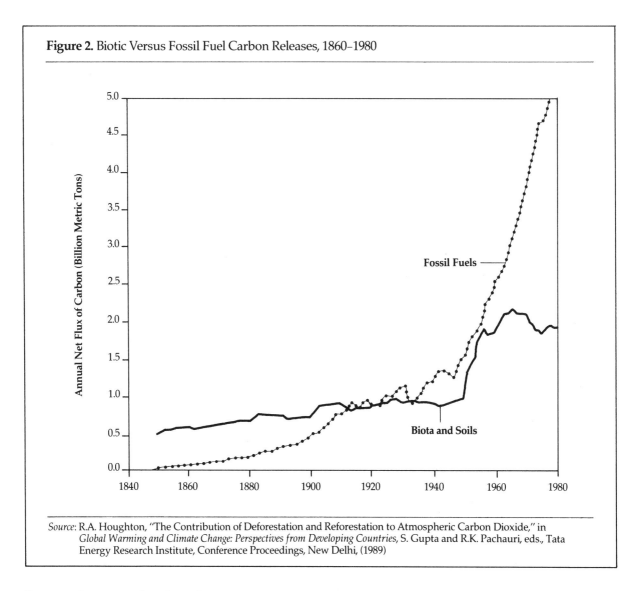

Figure 2. Biotic Versus Fossil Fuel Carbon Releases, 1860–1980

Fossil Fuels

Biota and Soils

Annual Net Flux of Carbon (Billion Metric Tons)

Source: R.A. Houghton, "The Contribution of Deforestation and Reforestation to Atmospheric Carbon Dioxide," in *Global Warming and Climate Change: Perspectives from Developing Countries*, S. Gupta and R.K. Pachauri, eds., Tata Energy Research Institute, Conference Proceedings, New Delhi, (1989)

In assessing general carbon flows in the United States, policymakers do have available to them considerable information on land-use trends, but this information constitutes just a small part of a complex picture. *(See Figure 5.)* U.S. forests are currently estimated to store some 57 billion tons of carbon. *(See Figure 6.)* Several key questions need answers, however, before the size and direction of net carbon flows can be estimated:

• How large are stocks of noncommercial biomass on timberlands? And how much carbon is stored in forested lands not classified as timberlands? What are the trends in each case?

• How much carbon is stored in wood products (both in use and disposed of)?

• How much carbon is stored in urban forests? How are trends in urban forestry affecting fossil fuel consumption?

• If the billions of tons of fossil fuel-emitted carbon previously thought to be disappearing

Figure 3. Distribution of United States Surface Area by Land Use

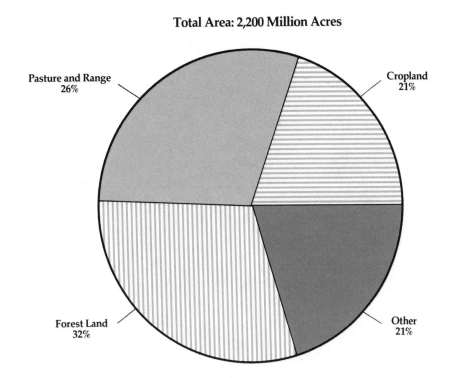

Total Area: 2,200 Million Acres

Pasture and Range
26%

Cropland
21%

Forest Land
32%

Other
21%

Source: H.T. Frey and R.W. Hexem, *Major Uses of Land in the United States: 1982*, Agricultural Economic Report No. 535, U.S. Department of Agriculture, Economic Research Service, (1985)

into the world's oceans every year are—as scientists now believe—instead fertilizing temperate biota around the world, what is the impact on domestic forests?[10]

These uncertainties notwithstanding, one key fact is incontrovertible. As hundreds of millions of acres of forests in what is now the United States were cut down over the last 200 years and converted to agriculture, pasture, and urban uses, and as virgin forests were replaced with younger forests and plantations, tens of billions of tons of carbon were added to the atmosphere. (For perspective, estimates of current annual global carbon releases through deforestation range between one and three billion tons.)

As hundreds of millions of acres of forests in what is now the United States were cut down over the last 200 years and converted to agriculture, pasture, and urban uses, and as virgin forests were replaced with younger forests and plantations, tens of billions of tons of carbon were added to the atmosphere.

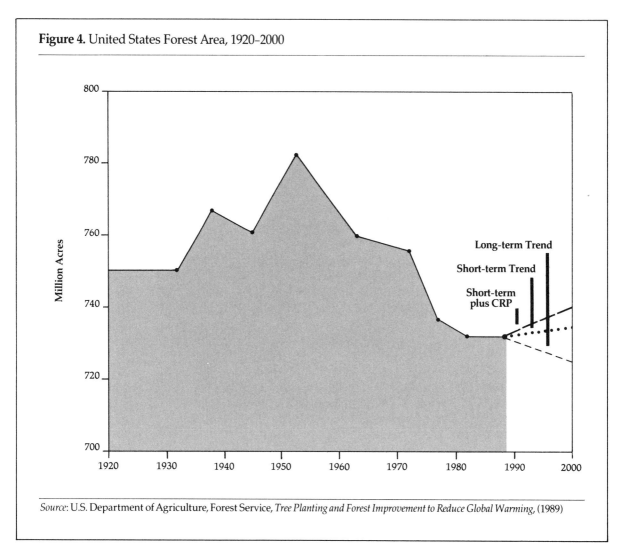

Figure 4. United States Forest Area, 1920–2000

Long-term Trend

Short-term Trend

Short-term plus CRP

Source: U.S. Department of Agriculture, Forest Service, *Tree Planting and Forest Improvement to Reduce Global Warming*, (1989)

Biotic carbon levels in what is now the United States have steadily declined since Europeans settled the continent, but this trend cannot last. Agricultural production has been concentrated on fewer and fewer acres, thus allowing the reversion of lands to forest cover. According to some modelers, this change has already resulted in U.S. forests' becoming net sinks for atmospheric CO_2. Richard Birdsey of the USFS concludes that timber growth now exceeds harvest by more than 120 million tons of carbon annually,[11] and Edward Hansen, also of the USFS, concludes that the size of the forest carbon reservoir is increasing by some 80 million tons per year.[12]

Such estimates are far from definitive, however. If annual nonharvested tree mortality is subtracted from total tree growth,[13] for instance, Birdsey's estimate of net carbon accumulation falls to just 11 million tons per year.[14] In addition, a variety of warning signs suggest a need for concern about future U.S. carbon stocks. *(See Box 4.)* A comprehensive model of biotic carbon flows in the United States might well show that the United States continues to be a net contributor of biotic carbon to the atmosphere. *(See Appendix.)*

A wild card that has thus far been largely overlooked in this analysis is global warming

Box 3. Global Warming and the Harvesting of Old-Growth Forest

Logging old-growth forests in the Pacific Northwest and Alaska has for years been the subject of dispute between logging and conservation interests. Logging interests argue that logging is vital to the economic health of the region. Opponents argue that the interests of threatened species and bio-diversity generally are left out in the U.S. Forest Service's (USFS) multiple-use calculus and that public subsidization of some of these sales cannot be justified. Because of its remote Alaskan location, for example, subsidies are necessary to make Tongass National Forest timber competitive with other Pacific Northwest national forests. The federal government is selling off 500-year-old spruce trees to loggers, and in some years has received just $0.10 for every tax dollar invested.

The debate over old-growth harvesting, particularly in the Tongass National Forest, has recently been cast in terms of its global warming implications. Senator Murkowski of Alaska, for example, has argued that 48 trees grow back for every old-growth tree removed, concluding that replacing some of the old forest with young trees increases the amount of CO_2 removed from the atmosphere.

In a recently released study, however, scientists report that the amount of carbon stored in Pacific Northwest old-growth is huge and that harvesting old-growth forest results in a net release of carbon dioxide to the atmosphere for at least two centuries. Harmon, Ferrell, and Franklin calculate that converting one acre of old-growth forest to younger forest reduces net terrestrial carbon storage by 135 tons even after the new trees have had a chance to grow for 60 years and

even if 42 percent of the originally harvested timber is assumed to be used for long-term product. The logging of some 13 million acres of Northwest old-growth forest since 1890, they therefore conclude, has added an estimated 1.6 to 2.0 billion tons of carbon to the atmosphere. They also estimate that only five million acres of such forest remain.

Although Alaskan old growth forest is believed to hold somewhat less biomass than old growth in the Pacific Northwest, it still stores significant amounts of carbon. In a study of Alaska's old growth, Paul Alaback of the USFS states that not only is tree biomass drastically reduced in a second-growth forest, but understory vegetation is also much lower. He concludes that replacing old-growth, whether in Alaska or the Pacific Northwest, adds significant quantities of carbon to the atmosphere over both the short and long terms.

Sources: Alaback, P.B., 1989. ''Logging of Temperate Rainforests and the Greenhouse Effect: Ecological Factors to Consider,'' in *Proceedings of Watershed '89: A Conference on the Stewardship of Soil, Air, and Water Resources.* U.S. Department of Agriculture, Forest Service; Harmon, M.E., Ferrell, W.K., and Franklin, J.F., 1990. ''Effects on Carbon Storage of Conversion of Old-Growth Forests to Young Forests,'' *Science* 247:699–702; Murkowski, F.H., 1989. ''The Tongass.'' *Christian Science Monitor*, March 14; Natural Resources Defense Council, 1989. *Cooling the Greenhouse: Vital First Steps to Combat Global Warming;* O'Toole, R., 1988. *Reforming the Forest Service.* Island Press; Rice, R.E., 1989. *National Forests: Policies for the Future, vol. 5, The Uncounted Costs of Logging.* Wilderness Society.

Figure 5. Important United States Biotic Carbon Flows

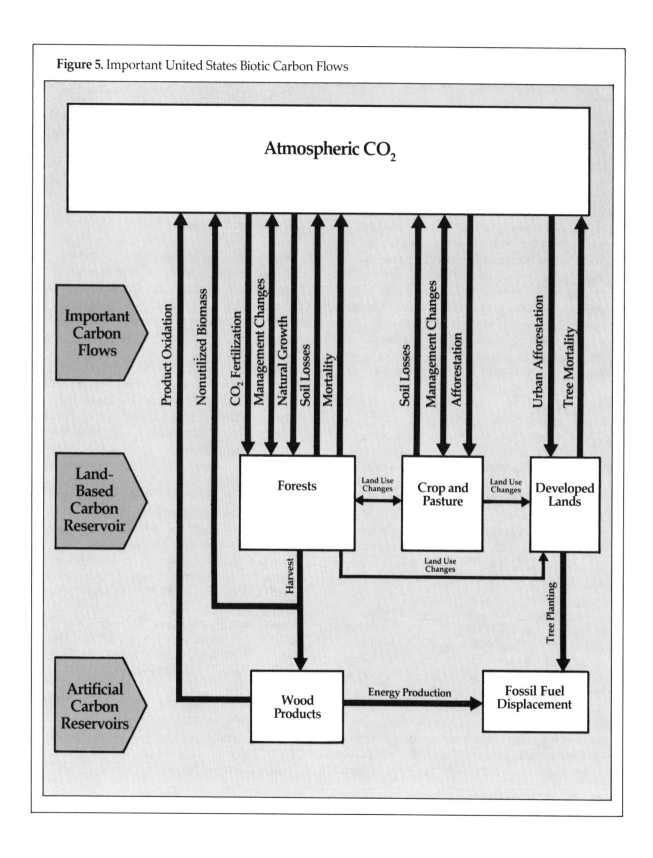

Figure 6. Carbon Content of United States Forest Ecosystems

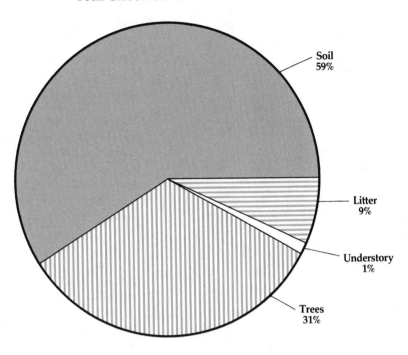

Total Carbon Content: 57 Billion Tons

Soil
59%

Litter
9%

Understory
1%

Trees
31%

Source: R.A. Birdsey, "Potential Changes in Carbon Storage Through Conversion of Lands to Plantation Forests," in
Proceeding, North American Conference on Forestry Responses to Climate Change, Washington, D.C. Climate Institute,
in press, (1990)

A comprehensive model of biotic carbon flows in the United States might well show that the United States continues to be a net contributor of biotic carbon to the atmosphere.

itself. If regional climates around the United States change significantly in coming decades, the impacts on forest resources could be considerable.[15] Although climate models differ in their projections of species-specific impacts, the equivalent of a doubling of atmospheric CO_2 levels could lead to climate change sufficient to shift the climatic ranges of tree species north by hundreds of miles, possibly leaving slowly maturing species behind to die out.[16] For established forests in the southern parts of species' current ranges, physical and biological stresses would be aggravated.[17] For some forests, water availability and soil moisture would decrease, intensifying natural fires.[18] More generally, as temperatures change, seedlings may be unable to regenerate, so much larger-scale artificial regeneration than has been required thus far might be needed. Insect pests and other pathogens moving up from points south could

Box 4. Timberland Carbon Stocks—Signs of an Uncertain Future

Statistics purporting to reflect the status of U.S. timberlands from one perspective or another are both voluminous and readily available. The forest industry and the U.S. Forest Service (USFS) can provide statistics showing that timberlands are doing well in the sense that commercial timber stocks continue to increase even while timberland acreage falls. Critics of current forest management practices can provide statistics showing the destruction of old-growth forests and localized declines in forest growth. Both interest groups are just now collecting the types of information that will ultimately indicate where net carbon flows are headed in the nation's timberlands. A variety of illuminating statistics, however, at least suggest the need for concern about trends affecting biotic carbon storage and emissions.

- Despite ongoing reforestation efforts, timberland acreage is projected to decline from the present level of 483 million acres to 475 million acres by 2000 and to 462 million acres by 2040. Most of these losses will be from private timberlands, and they are real—not just reclassification casualties.

- Standing merchantable softwood timber volumes in national forests are projected to decline by almost 30 percent between 1986 and 2000, or some 60 billion cubic feet. As a result, total softwood carbon inventories on national forest lands will probably fall by more than one billion tons by the year 2000. If soil carbon

losses are added in, the figure will probably be higher.

- Although private timberland holdings account for most timber production in the United States, it is hard to predict how changing management practices on these lands might affect carbon stores. Anecdotal evidence suggests, for example, that timber companies nationwide are already stepping up the conversion of long-term timber stands. Pacific Lumber Co. is a former merger target that is now saddled with heavy debt. Since it was taken over in 1985 by financier Charles Hurwitz, the company has doubled or possibly even tripled its rate of logging in its redwood stands, which comprise three-fourths of the privately owned uncut redwoods in the United States. The company now estimates that these stands will disappear within 25 years; outside observers say 3 to 10 years.

- Growth rates in southern pine plantations are declining for still poorly understood reasons. If plantation growth in the southern United States ultimately falls by 5 to 10 percent, as evaluated in USFS scenarios, it would lead to a decline in timber inventories in the Southeast of more than 20 percent by 2040.

- USFS statistics suggest that harvesting areas containing old-growth forest will continue at high rates, bringing standing inventory down by 25 percent between

increase tree mortality significantly, and rising temperatures and more frequent drought could aggravate forest damage from pollutants such as ozone. Although U.S. forests would probably

reach a new equilibrium eventually if climate did so, the many transitional years would be stressful ones that would likely sap general forest health, alter growth rates, lower lumber

1986 and the year 2000. In some forests, harvest rates are more than double forest growth. As much as two billion tons of carbon have already been released on net through old-growth harvesting during the past century in the Pacific Northwest; hundreds of millions of additional tons will be released if these old-growth forests are ultimately eliminated.

- After already almost doubling between 1952 and 1986, the demand for commercial timber from U.S. forests is projected to increase by 50 percent between 1986 and 2040. Total timber consumption in the United States is projected to increase from 20.5 billion cubic feet in 1986 to 28.6 billion cubic feet by 2040. How much increasing timber consumption might increase the stocks of carbon stored in timber products remains quite uncertain.

- Annual harvest levels on U.S. timberlands are expected to rise over 25 percent to some 5.4 million acres by 2030. Remaining natural pine stands will decline in area by about 50 percent, or 20 million acres, and will be replaced with pine plantations. These changes could have either of two basic effects on carbon stocks. On the one hand, growth and stocking rates might increase, thus increasing carbon stores. On the other, the carbon stored in noncommercial timber and in soils could decrease.

- Natural disasters in recent years have destroyed considerable quantities of timber, and drought conditions in many

parts of the United States are a serious threat to timber resources. More than five million acres of U.S. forest were seriously damaged or destroyed by fire in 1988. In the same year, Midwest droughts raised the mortality rate of seedlings to between 60 and 80 percent, much higher than normal. About 250 million seedlings on 300,000 acres of forest plantations were lost nationally. In 1989, hurricane Hugo damaged or destroyed two to five million acres of trees in the Carolinas.

- Environmental concerns are challenging timber-management practices in ways that may reduce growth rates, particularly on public lands. It is estimated, for example, that 24 percent of the lands replanted by the Bureau of Land Management between spring 1986 and spring 1989 need planting again, primarily because of a court-imposed prohibition on herbicide use in the planted areas.

Sources: U.S. Department of Agriculture, Forest Service, 1990. *An Analysis of the Timber Situation in the United States: 1989–2040.* Editorial draft; Ryan, J.C., 1989. ''Wall Street Goes Wild,'' *World Watch* 2(6):7; President's Interagency Drought Policy Committee, 1988. *The Drought of 1988, Final Report of the President's Interagency Drought Policy Committee;* Lewis, R., Bureau of Land Management, Medford District, pers. comm. to the Office of Technology Assessment, January 22, 1990; U.S. Department of Agriculture, Forest Service, 1989. *RPA Assessment of the Forest and Range Land Situation in the United States, 1989.*

quality, and undermine wildlife habitats. Major changes in carbon flows and reservoirs might result, almost certainly increasing net emissions of carbon, potentially seriously.

Conclusions

Carbon dioxide emissions from human activities account for just a small fraction of the roughly 210 billion tons of carbon that cycle annually between the atmosphere and terrestrial biota.[19] But because it is these activities that are responsible for the unwanted buildup of CO_2 in the atmosphere, modifying them should help moderate the rate and ultimate magnitude of global warming. Reducing fossil fuel CO_2 emissions in the first place is the most obvious place to start, but changing land-use practices and patterns can play a role as well. Five land-use-based approaches can be used to slow the buildup of CO_2 in the atmosphere:

- slowing or stopping the loss of existing forests, thus preserving current carbon reservoirs;

- adding to the planet's vegetative cover through reforestation or other means, thus enlarging living terrestrial carbon reservoirs;

- increasing the carbon stored in nonliving carbon reservoirs such as agricultural soils;

- increasing the carbon stored in artificial reservoirs, including timber products; and

- substituting sustainable biomass energy sources for fossil fuel consumption, thus reducing energy-related carbon emissions.

Because biotic policy options appear capable of contributing significantly to the mitigation of global warming while also furthering many other public policy objectives, their role deserves careful consideration on a country-by-country basis.

These approaches are all based on the same basic premise: adding to the planet's net carbon stores in vegetative cover or soil, or preventing any net loss, will help moderate global warming by keeping atmospheric CO_2 levels lower than they would otherwise be. Biotic carbon emissions are a significant part of total anthropogenic carbon emissions, exceeding fossil fuel emissions in many countries. Because biotic policy options appear capable of contributing significantly to the mitigation of global warming while also furthering many other public policy objectives, their role deserves careful consideration on a country-by-country basis.

III. Slowing Global Warming Biotically—Options for the United States

Whether deploying biotic options will slow the accumulation of CO_2 in the atmosphere depends on the type and quantity of vegetation being added to existing stocks or being protected from loss. It also depends on what eventually happens to the biomass thus added or protected. Planted on what would otherwise be bare land, even a vegetative cover that grows in summer and decomposes completely in winter will lower average annual CO_2 concentrations. Planting perennial vegetation on bare ground, however, lowers atmospheric CO_2 concentrations by the increment of carbon stored in the new perennial biomass and added to soils. Perennial vegetation normally accumulates much more carbon per acre over time than annual vegetation can, and thus it plays a far greater role in expanding terrestrial carbon reservoirs over the long term.

It makes no difference to atmospheric CO_2 concentrations whether a ton of carbon is added to the vegetation of a temperate or a tropical country. The applicability of the different biotic approaches, however, is likely to vary widely from country to country. Slowing deforestation, for example, would obviously prove most beneficial in tropical countries where significant deforestation is taking place. The best way to expand vegetative carbon reservoirs might be new plantations in one country and better management of existing forests in another. And although substituting biomass fuels for fossil fuels has significant potential throughout the world, the technological infrastructure for taking advantage of the option in the short term is generally better developed in temperate countries.

It makes no difference to atmospheric CO_2 concentrations whether a ton of carbon is added to the vegetation of a temperate or a tropical country.

Each of the five biotic approaches introduced in Chapter II is applicable to some extent in the United States. Taking U.S. land-use characteristics into account, Table 1 presents a menu of policy options tailored to U.S. carbon storage opportunities. Several of the options are capable of significantly reducing net U.S. carbon emissions; several offer corollary benefits in areas other than global warming mitigation. The time frame and costs of the different options vary widely, although in most cases some level of implementation appears economically justified even without considering global warming. *(See Box 5.)* The approach, projected costs, and advantages of seven different policy options are profiled below.

Table 1. Options for Increasing the Uptake and Utilization of Biotic Carbon in the United States

—Planting Trees to Reduce Energy Use

—Converting Private Agricultural and Pasture Lands to Conventionally Managed Tree Cover

—Increasing Tree Cover on Nonforested Public Lands

—Stocking and Managing Already Forested Lands More Intensively

—Substituting Wood for Other Products

—Increasing the Commercial Use of Biomass Energy

—Improving Soil-Management Practices

1. Planting Trees to Reduce Urban Energy Use

Expanding and better managing our urban forests is commonly discussed as one of the most effective biotic ways to combat global warming in the United States. It is the primary focus of the American Forestry Association's (AFA) recently launched Global Releaf program, which calls for planting 100 million urban trees by 1992. It is also an important focus of the America the Beautiful initiative proposed in President Bush's 1991 budget, which calls for planting 30 million trees per year in U.S. communities. *(See Box 6.)*

Urban trees, and to some extent other types of vegetation, can slow the accumulation of CO_2 in the atmosphere by reducing urban energy demand. They can shade houses and small buildings from the sun and transpire large quantities of water,[20] thus lowering

ambient temperatures and reducing cooling loads. They can also increase the reflectivity of dark asphalt surfaces and other elements of urban environments, thus reducing the amount of solar radiation absorbed and reemitted as heat.[21] And during winter, vegetation can help insulate structures from cold winds. For these reasons, an urban tree can be up to 15 times as effective at "reducing" atmospheric CO_2 as a rural tree.[22]

An urban tree can be up to 15 times as effective at "reducing" atmospheric CO_2 as a rural tree.

Researchers at Lawrence Berkeley Laboratory have done extensive modeling of the relative benefits of placing trees around residences and small buildings in climate zones where air conditioning is used.[23] Random placement of three trees around an otherwise unshaded house yielded reductions in annual cooling energy requirements of 11 to 42 percent; the strategic placement of the three trees on the southern and western exposures of the house reduced cooling energy requirements by 13 to 52 percent. Of the various energy impacts of urban trees, so-called evapotranspiration is the most important.[24] In this process, trees are natural air conditioners, alleviating some of the heat-island effect associated with human activities and the builtup environment in urban areas. In an urban setting, neighborhoods with mature tree cover record daytime peak temperatures 3° to 5° F lower than in newer areas with no trees.[25] In Washington, D.C., each 1°F temperature rise during hot summer days increases area cooling costs by $10,000 per hour. Nationally, the urban heat-island effect costs an estimated $1 million per degree per hour.[26]

How much energy urban trees save during winter is less well investigated. It is known that trees help reduce outside air infiltration

Box 5. Quantifying the Costs of Mitigating Global Warming Biotically

Estimating the life-cycle costs (or savings) of purchasing a more efficient electrical appliance or a more efficient light bulb is quite straightforward. The up-front purchase price of new appliances or the light bulbs is compared, as is the present value of the stream of energy payments associated with their operation. If the life-cycle cost of the more efficient appliance or bulb is lower than that of the alternative, the cost of sequestering carbon through energy efficiency is virtually zero.

Unfortunately, assessing the life-cycle costs of implementing biotic policy options is far more difficult than determining the life-cycle costs of a light bulb. In the case of converting private lands, for example, the various opportunity costs confronting landowners are much more complex and difficult to estimate than those facing buyers of electric appliances. The National Research Council estimates that land-use decisions on over 60 percent of harvested croplands are heavily influenced by federal program rules and incentives and that such programs will continue to shape land-use decisions in the future. And even though total federal outlays for direct payments to farmers, export subsidies, storage costs, and loans have been declining since 1986, partially as a result of drought-related increases in many commodity prices, they still accounted for 40 percent of net farm income in 1987. Historically, the net effect of agricultural incentives has been to increase agricultural cultivation on lands ill suited to farming. Although the 1985 Farm Bill has alleviated this problem somewhat, farm programs continue to be complex and sometimes contradictory. Many

impose obstacles to adopting agricultural and land-use policies that would encourage the conservation and buildup of soil carbon, the reduction of energy inputs, and the withdrawal of unsuitable lands from agricultural production. Even though nearly 70 million acres were idled in set-asides in 1987, total agricultural capacity was still some 16 percent in excess of what prevailing market prices would justify.

Apart from the difficulties of assessing the true societal costs of changing land uses to mitigate global warming, other variables complicate the process further. The initial costs of forestry options can vary widely, as can the recurring costs of managing trees. The carbon benefits of tree planting are themselves subject to uncertainties that policy implementors cannot control—among them, long-term biomass growth rates and natural disasters. And further, the conventional economic payoff of planting trees is so far away, often 50 to 75 years, that it is often ignored in estimating the cost-effectiveness of biotic options. For these reasons, the costs of sequestering or displacing carbon through biotic options can vary so much that comparing the apparent cost-effectiveness of such options to each other and to other mitigation options (such as appliance efficiency) can be frustratingly difficult.

Sources: National Research Council, 1989. *Alternative Agriculture.* National Academy Press; U.S. General Accounting Office, 1990. *Alternative Agriculture: Federal Incentives and Farmers' Opinions.* Program Evaluation and Methodology Division.

Box 6. The America the Beautiful Initiative

As part of its proposed fiscal year 1991 budget, the Bush administration included an initiative known as America the Beautiful, an important element of which is the annual planting of one billion trees as part of a national reforestation program aimed at slowing global warming. Under this program, $175 million was tentatively budgeted for the 1991 fiscal year to promote both rural and urban tree planting. Of this, $110 million is for reforesting 1.5 million acres per year and initiating timber-stand improvement practices on an additional 180,000 acres. The remaining $65 million is earmarked for the proposed urban forestry component of the program—planting 30 million trees annually in 40,000 communities around the country. Of the total set-aside for urban forestry, $30 million is for coordination and technical assistance, and there is an additional one-time allocation of $35 million for capitalizing a private nonprofit foundation to lend nongovernmental leadership to the effort and build further financial support. Overall leadership of the reforestation efforts would be shared by the Forest Service and the new community forestry foundation.

The carbon benefits of the rural reforestation component of the America the Beautiful program can be calculated quite easily if simplifying assumptions are made. If an additional 1.5 million acres of trees are actually planted per year as a result of the new program, if this move does not prompt other would-be tree farmers to forgo planting or reforesting their own lands, and if no natural or man-made problems (such as fire or pollution) keep the trees from surviving and growing, calculating the carbon benefit of the program is simply a matter of selecting a growth rate. Using the average productivity of all U.S. commercial forests, or about 0.6 tons of carbon per acre per year, the pro-

gram would draw about 900,000 tons of carbon per year from the atmosphere. If a higher growth rate is assumed to result from aggressive plantation management, perhaps as much as 2.5 tons of carbon per acre per year, annual carbon accumulation could reach 3.75 million tons. If the planting were continued for 10 years, the 15 million acres of trees planted could withdraw between 9.0 and 37.5 million tons of carbon from the atmosphere per year for several decades, roughly the equivalent of 0.5 to 2.3 percent of current U.S. fossil fuel emissions.

Calculating the carbon benefits of the urban forestry component of the America the Beautiful program is much more difficult, because some significant fraction of the trees is apparently intended for communities where cooling energy requirements are low. Planting trees in such cities would probably release more carbon than it would absorb or displace. Where space-cooling requirements are high, however, the carbon content of fossil fuel consumption eventually forgone through the program after 10 years of planting might be as high as 5 million tons per year, making the total program's carbon benefits the equivalent of 0.8 to 2.6 percent of current U.S. fossil fuel emissions.

The economics of the America the Beautiful program, however, are questionable. Even with optimistic assumptions, the $110 million targeted for rural forestry allows a one-time payment of less than $60 per acre to convince farmers to convert their unforested land to tree-crop cover. Although EPA and Department of Agriculture (USDA) analyses underlying the proposal suggested that this would likely prove insufficient, and indeed that rental payments of the sort seen in the CRP program would become necessary relatively rapidly, the Office of Management and Budget reportedly rejected their conclusions. Similarly, the $30 million budgeted for urban tree planting will cover only a small fraction of the total costs associated with planting 30 million new trees per year.

Although early reactions from the nongovernmental forestry community to the America the Beautiful program have been positive, other line items in the overall forestry budget are controversial. Many proposals for reducing or eliminating existing programs in the forestry sector will negate some of the benefits being attributed to the America the Beautiful program. Problems that have been identified by forestry groups include:

• A reduction in other reforestation funding by more than $4.5 million, notwithstanding an increase in area to be reforested.

• A reduction of spending on state and private forestry from almost $100 to $56.6 million.

• The elimination of funding for insect and disease suppression on cooperative state and private lands, which currently stands at $10.6 million.

• A reduction in the firefighting budget for cooperative state and private lands from $14 to $4 million.

• A reduction by almost 50 percent in the forest management and utilization budget.

• The elimination of the federal forest stewardship and rural development initiative funding. These programs currently promote environmentally sound multiple uses of land, including timber production on nonindustrial private forest lands. The program is proposed for elimination just as states are completing state stewardship plans intended to bring 25 million acres of nonindustrial private forest under stewardship management over the next five years.

• Reductions in the nursery and tree improvement program and urban forestry program funding.

• Elimination of the tropical forestry initiative, currently at $4 million per year.

• The failure to repay the USFS's Knudsen Vandenberg account for some $244 million in firefighting expenses beyond what was budgeted for in past years. The proposed budget for forest firefighting is $118 million, though costs over the last three years have averaged more than $300 million per year. The continued siphoning off of money to pay for firefighting may detract from the funding of other forestry activities in coming years.

• The likely closure of five additional USFS research stations, an increase in the interval between timber inventories from 10 to 13 years, the elimination of funding for remote sensing, and the reduction of funding for international trade research.

Many experts contend that many of these programs should be expanded, not reduced or eliminated. Expansion is needed, they say, to improve the health of U.S. forests as well as to help prepare those forests for the changes that may result from global warming and other environmental changes.

Sources: U.S. Department of Agriculture, Forest Service, 1990. *America the Beautiful: National Tree Planting Initiative;* Dinus, R.J., Francis, M.S., Sampson, N., and Mixon, J., 1990. Statements before the House Agricultural Committee, Forest, Family Farms, and Energy Subcommittee, February 6.

into buildings by making the city surface rougher and lowering wind speeds. According to one report, under certain conditions windbreaks around homes and buildings can reduce heating energy use by 10 to 50 percent.[27]

The Potential Magnitude of Urban Tree-Planting Benefits. The AFA estimates that the average urban area has a 30 percent canopy cover, which is only half of such areas' potential capacity. President Bush's America the Beautiful proposal to plant as many as 600 million new trees in urban areas over 20 years would virtually double this canopy. The AFA's more modest goal reflects its belief that there are 100 million sites for new trees around homes and small businesses that would yield particularly significant energy savings.[28]

According to Lawrence Berkeley Laboratory, 100 million new trees could—if properly placed—prevent the emission through fossil fuel combustion of approximately 10 million tons of carbon by reducing national cooling energy requirements by 10 percent.[29] To realize such gains, the new urban trees would have to be planted around one-half the air-conditioned homes and small commercial buildings in the United States. If the reflectivity of these buildings were also increased—by, for instance, using more reflective roofing materials—the energy savings could be almost doubled, thus reducing carbon emissions by some 20 million tons. Total coverage of available planting opportunities would in principle result in even greater reductions, perhaps as high as 15 million tons.

From a carbon mitigation perspective, the maximum achievable benefits of urban tree-planting programs are modest. Only 2.7 quads of energy, or less than four percent of U.S. annual energy consumption, are used to cool residences and buildings (1.2 and 1.5 quads, respectively). Although producing this much energy emits some 50 million tons of carbon, much of it is consumed in large commercial buildings and other settings where trees do relatively little good. In addition, several factors will affect the timing and realization of carbon savings projected for large-scale urban tree planting. First, trees need considerable time to grow large enough to shade buildings and transpire enough water to cool the local environment. They must also be properly placed around the house or building, and they must be kept healthy. Second, space-cooling technologies and buildings are themselves becoming more energy efficient over time: replacing an old residential air conditioner with one that is 50 percent more efficient will cut by about one-third the potential carbon benefits associated with planting trees around the house. Third, urban tree planting and maintenance consume energy in their own right. The energy used to grow, plant, water, fertilize, prune, spray, and otherwise maintain trees in urban areas where little or no air conditioning is used might well release more carbon than the trees take up or displace. Leaf collection and dead tree removal also consume significant amounts of energy.

From a carbon mitigation perspective, the maximum achievable benefits of urban tree-planting programs are modest.

The Costs of Urban Tree Planting as a Mitigation Strategy. Simply handing out tiny seedlings at fast-food chains or other establishments costs very little per tree, but a high percentage of these seedlings is likely to perish. Many more will not be planted where they will maximize energy savings. As a result, the "cheapest" trees are likely to be less reliable in mitigating CO_2 emissions, and they may prove less cost-effective than trees planted through sophisticated programs with much higher upfront costs. In one such scenario, with a one-time cost of trees of $15 to $75 per home (compared to a recurring expenditure of $100 per year for air conditioning), the estimated annualized cost per ton of carbon saved is $6 to $26.[30] However, providing needed extension services to ensure that trees are placed properly, finding the personnel and providing the infrastructures needed to maintain the trees, as well as countering the problems already facing urban forests of pollution and numerous other stresses, would all add to the estimated costs. So would relying on larger trees in the hope of speeding up the accrual of significant benefits, which would otherwise take well over a decade to discern.

The Pros and Cons of Urban Tree Planting. Compared to other types of forestry, urban tree planting has a number of advantages. It can, like the AFA's Global Releaf program, mobilize considerable citizen participation, thereby sensitizing the public to global warming and other environmental concerns. It can serve aesthetic needs and help advance such urban environmental goals as slowing the movement of storm water, thereby reducing erosion and combatting sewer overload. Even more important, this approach can decrease the need for expensive new powerplant capacity if urban air-conditioning demand at peak periods can be constrained—probably an even more significant benefit than its effect on slowing global warming.

These benefits notwithstanding, many barriers will impede the long-term success of an urban tree-planting program intended to mitigate global warming or reduce peak energy demand. More energy might well be consumed in maintaining improperly placed trees than is saved by them, especially if they reduce solar gain during the winter and thus increase heating costs. From policymakers to homeowners, the range of people who must be educated and mobilized to plant trees correctly, then maintain and protect them, can be surprisingly large. In addition, urban trees are already beleaguered by air pollution, compacted soils, inadequate growing pits and water, such irritants as salt, careless road and municipal repairs, and many other problems. Today, the average downtown tree survives just seven years; even suburban trees last only an estimated 32 years. A 1986 survey of 20 larger cities found that four trees are dying or dead for every one tree being planted; in New York and Chicago, the ratio rises to eight to one.[31] Ironically, trees themselves contribute to the production of volatile organic hydrocarbons, one of the precursors to the urban smog that poisons trees.[32]

From policymakers to homeowners, the range of people who must be educated and mobilized to plant trees correctly, then maintain and protect them, can be surprisingly large.

Past efforts to increase the number of urban trees have met with varied success. One million seedlings were distributed for planting in Los Angeles before the 1984 Olympics; most apparently died. Whether the programs now starting up will turn the tide will take years to determine. Given the many barriers to overcome in expanding urban forests, however, it is unlikely that we can come close to reducing carbon emissions by the 15 million tons suggested as technically feasible. As a practical matter, even a broad-based policy initiative in the urban forestry sector appears unlikely to displace U.S. fossil fuel emissions by more than 20 to 30 percent of the technical potential. Indeed, just stopping the continuing deterioration of urban trees and the resulting intensification of the urban heat-island effect will take considerable effort. Urban forestry budgets and staffing have declined markedly over the last decade and the $500 million currently spent annually on municipal tree care is a far cry from the amount required even to maintain our urban forests properly, much less to expand them.[33]

Stopping the continuing deterioration of urban trees and the resulting intensification of the urban heat-island effect will take considerable effort.

As this brief analysis shows, there are many worthwhile reasons to plant and protect trees in urban areas. As a carbon storage and displacement option, tree planting could in principle provide a carbon benefit of some 15 million tons, although 3 to 5 million tons seems more likely even with an aggressive policy effort. In regions that have high cooling-energy demands and high energy costs and that face the construction of expensive additional electrical generating capacity, the cost of this policy option on a dollars-per-ton basis could be very low. Costs increase as cooling demand or capacity value falls, as extension and educational efforts become more sophisticated, and as larger trees are used to accelerate the accrual of benefits. After an increment of almost free carbon storage, most of the practically achievable carbon benefits might well cost $10 to $30 per ton. These costs could rise much higher in areas with modest space-cooling requirements or in communities with already relatively full tree canopies.

2. Converting Private Agricultural and Pasture Lands to Conventionally Managed Tree Cover

Some 250 million of the approximately 1 billion acres currently in agricultural, pasture, or rangeland uses were forested before Europeans settled the continent. Many of these once-forested acres are highly erodible or otherwise environmentally fragile, but they remain in agricultural production or related uses. The same can be said for millions of additional acres that could, in principle, support trees or other crops for carbon storage or energy production.

One way to convert more private lands to tree cover is to expand existing programs such as the Agricultural Conservation Program (ACP) and Forestry Incentives Program (FIP). *(See Box 7.)* Alternatively, the acreage goals of the Conservation Reserve Program (CRP) could be expanded and its criteria loosened to encourage the conversion of additional acreage to tree cover; in addition, incentives could be adopted for converting to tree-cover acreage already signed into the CRP. The goal of increasing forested acreage might also justify a new program, possibly termed the Forestry Reserve Program (FRP), that would function much as the CRP does but without the same eligibility criteria. Because the pool of eligible land could include agricultural land of considerably lower value than that now targeted for CRP inclusion, government payments could be kept lower with this approach.

A related land-conversion option is to encourage the construction of windbreaks around agricultural fields, on rural homesteads, and along rural roads. Such windbreaks can store carbon, reduce fossil fuel consumption, and provide ancillary economic and environmental benefits, such as reduced soil erosion.

Other opportunities also exist for converting agricultural land to further carbon-storage goals at least moderately. Promoting research, development, and market penetration for commodities such as kenaf—which yields a fiber that can be used to make paper—could free up land now used to produce wood pulp to produce wood for more long-lived products and could ultimately take some harvesting pressure off U.S. timber resources.[34]

The Potential Magnitude of Carbon-Storage Benefits. Some 328 million acres were used to grow crops during 1988.[35] Another nearly 80 million acres were idled to reduce production or conserve land, an additional 65 million were used temporarily as pasture in cropland-rotation schemes, and almost 600 million acres were in permanent pasture or rangeland.

How much land might reasonably be converted to tree cover? Estimates vary widely depending on the criteria used. For the nation as a whole, the U.S. Department of Agriculture (USDA) estimates that some 310 million nonforested acres are capable of supporting tree growth.[36] Estimates of the quantity of economically marginal or environmentally sensitive crop and pasture lands that could physically support tree cover range from about 120 million acres to almost 250 million acres.[37] In the eastern United States alone, some 40.4 million acres of cropland and 41.7 million acres of pasture land are estimated to fall into these two categories.[38] Some 26 million acres, most of them in the southeastern United States, would already be more profitable to their owners if planted in trees even without changes in governmental or private incentives.[39]

The carbon-storage potential and cost-effectiveness of converting environmentally sensitive lands to tree cover depends to a large extent on how fast the trees grow. Current annual productivity in U.S. forests ranges widely, from 0.6 tons of carbon per acre for the average timberland to more than 2 tons of carbon per acre for intensive plantations.[40] Production targets for some species range as high as six to eight tons of carbon uptake per acre per year under particularly intensive management. But for the foreseeable future, an acre of land converted to a mid- to long-rotation tree-crop cover is more likely to incorporate from 1 to 2.5 tons of carbon per year into new biomass.[41] Significant silvicultural advances would be needed to reach the 2.5-ton level; Richard Birdsey of the USFS believes a figure of 1.5 tons of carbon per acre (including soil carbon) is realistic.[42] But keeping an upper limit of 2.5 tons per acre, to be optimistic, planting an additional acre of trees could result in 50 to 125 tons of carbon storage per acre over a 50-year growth cycle and for as long as the trees remain standing.

Unfortunately, the actual long-term carbon benefit of planting this acre of trees cannot be determined simply by extrapolating from projected annual biomass growth rates. First, the fossil fuel energy inputs needed to plant and maintain the trees must be subtracted from net biomass accumulation. In general, more energy is needed the more intensively one manages a tree crop and the higher the growth rates achieved. Net carbon benefits per acre also depend on whether the timber is eventually harvested and on what happens to both the merchantable and nonmerchantable portions of the harvested biomass.[43] When an acre of trees is harvested, a significant proportion of the carbon accumulated in biomass and in the soils can be lost through oxidation to the atmosphere during or soon after harvest. The extent to which carbon-storage benefits extend past the point of harvesting depends on the proportion of harvested carbon that is added to product stocks, the amount of fossil fuel energy displaced by the combustion of nonmerchantable wood and wastes, the amount of soil carbon loss, and the extent and speed of replanting and regrowth on the acre. Given these factors, an acre of new trees evaluated on the basis of net carbon storage within a decade of harvesting could easily yield considerably less than the 50 to 125 tons of carbon benefit calculated for an unharvested acre.[44] Indeed, even 35 to 90 tons of carbon per acre is probably an optimistic estimate.

When an acre of trees is harvested, a significant proportion of the carbon accumulated in biomass and in the soils can be lost through oxidation to the atmosphere during or soon after harvest.

The technical potential of shelterbelt options to mitigate global warming is difficult to assess. The equivalent of more than 3 million acres of windbreaks would be required just to protect

Box 7. Past and Current U.S. Reforestation Programs

Although global warming has not been a motivating factor in promoting domestic forestry until recently, large-scale reforestation has been promoted for other policy reasons for decades. The five major initiatives have been:

Civilian Conservation Corps: Over nine years in the 1930s and early 1940s, the Civilian Conservation Corps (CCC) planted 2.3 million acres of trees. In addition, the CCC practiced conservation efforts to constrain soil erosion, protect against forest fire, and improve wildlife habitat. At the program's peak, there were 311 forestry camps in the South alone on both national forest and private land. With the onset of World War II, tree planting and fire-protection efforts declined because the workforce was needed elsewhere; the CCC was disbanded in 1942.

Soil Bank: Between 1956 and 1962, some 2.2 million acres of trees were planted as part of the Soil Bank program; 1.9 million acres of them were planted in the South, contributing significantly to that region's timber supplies. The program was enacted to conserve soil and to reduce surplus farm production. The program paid for tree-planting costs and made annual cash payments to the participating landowners for up to 10 years, much as the CRP does today. Follow-up studies have shown that most of the land planted in trees stayed out of crop production, even after cash payments were discontinued; according to one study, 86 percent of Soil Bank plantations in the South were still in trees in 1980.

Agricultural Conservation Program: Created as part of the Soil Conservation and Domestic Allotment Act of 1936, the Agricultural Conservation Program (ACP) has focused on promoting farm-conserving practices. Like other conservation programs, its work has included tree planting, timber-stand improvement, and wildlife habitat enhancement. Federal cost sharing of tree-planting expenses under the program encouraged annual reforestation of 200,000 to 300,000 acres in the early 1960s, but that number has fallen to less than 100,000 acres per year in the last 15 years. Timber stands on an additional 200,000 to 250,000 acres per year were improved under the ACP in the early 1960s, but by 1986 that figure too had dropped—to some 27,000 acres per year. The government's share of costs has averaged $60 to $75 per acre (real 1988 dollars), with the total spent on reforestation and timber-stand improvement falling from approximately $24 million in 1960 to $7 million in 1986 (real 1988 dollars).

Forestry Incentives Program: Added as a rider to the Agriculture and Consumer Protection Act of 1973, the Forestry Incentives Program was created in response to the dwindling availability of ACP funding. In addition to cost-share payments for tree planting and stand improvement, the FIP also pays for site preparation for natural regeneration and for firebreak construction. Eligibility is restricted to nonindustrial private forest landowners who are not primarily involved with manufacturing forest products; in addition, land must be classed as commercial timberland and tracts of land must be between 10 and 10,000 acres. Since 1973, an average of 150,000 to 200,000 acres per year have been reforested under the program; timber-stand improvement has fallen from an average of 150,000 acres annually in the program's early years to approximately 40,000 acres during the past several years. Federal costs in 1988 dollars have fallen as well, from nearly $23 million in 1974 to less than $12 million in 1986.

The Conservation Reserve Program: Many acres of grasslands and wetlands were converted to agricultural use during the 1970s due to high food prices and federal crop-price-support programs. Many of these acres have proven quite fragile, however, and erosion is severe. And at today's food and land prices, almost 100 million acres of U.S. cropland are in excess of demand. The CRP was established as part of the Food Security Act of 1985 to foster the withdrawal from agricultural production of between 40 and 45 million of these highly erodible acres by 1990. To be eligible for inclusion in the CRP, land must have an average erosion rate of at least 19.1 tons of soil per acre—nearly three times the national average. By this criterion, an estimated 24 percent of all U.S. cropland—mostly in the Corn Belt, northern and southern plains, and mountain regions—is potentially eligible. Without special dispensation, however, CRP sign-ups are limited to 25 percent of a county's cropland in an attempt to spare local economies significant damage. This limitation reduces eligible land by almost one-third, to just under 70 million acres. To encourage landowners to participate in the CRP, the government shares the cost of converting the land to an alternative use and provides rental payments throughout the 10-year CRP contracts.

The erosion-control element of the CRP appears successful. Erosion on acreage signed into the program fell from an annual estimated average of 20.9 tons per acre before enrollment to 1.6 tons per acre after enrollment and the implementation of conservation treatments. Overall, CRP implementation through 1989 is estimated to be reducing total erosion by more than 600 million tons per year. CRP's reforestation element, however, has proven less successful. Congress hoped that 5.6 million acres of CRP enrollments would be converted to tree cover. With 33.9 million acres of land enrolled in the CRP as of early 1990, only 2.2 million acres were contracted to go into tree cover. (Tree planting has been largely concentrated in Georgia, Mississippi, Alabama, and South Carolina.) Some analysts believe, however, that some proportion—possibly as much as 650,000 acres—of CRP land currently in grasslands ultimately will be converted to tree cover by landowners.

In a recent survey of the obstacles to planting trees on CRP land, the reason farmers most often gave was the physical unsuitability of their land for growing trees, particularly owing to arid climate or sandy soils. Other common responses included beliefs that trees would interfere with the use of land after the 10-year CRP contract expired and that high maintenance costs and the long wait before trees could be harvested would make the switch unprofitable. The study also found, however, that farmers were more likely to plant trees if they had regular contact with the state forestry or agricultural extension agencies as well as prior commercial experience with planting or harvesting trees.

Sources: Cubbage, F.W., forthcoming. *Current Federal Land Conversion Programs: Accomplishments, Effectiveness, and Efficiency.* American Forestry Association; U.S. Department of Agriculture, Forest Service, 1988. *The South's Fourth Forest: Alternatives for the Future;* Esseks, J.D., Kraft, S.E., and Moulton, R.J., forthcoming. *A Survey Research Perspective on the Forestry Component of the Conservation Reserve Program.* American Forestry Association; Moulton, R.J., and Dicks, M.R., 1987. "Implications of the 1985 Farm Act for Forestry," in *Proceedings of the 1987 Joint Meeting of Southern Forest Economists and Midwest Forest Economists.* 1987, Asheville, North Carolina; U.S. Department of Agriculture, 1989. *Conservation Reserve Program: Progress Report and Preliminary Evaluation of the First Two Years;* U.S. Department of Agriculture, Agricultural Stabilization and Conservation Service, 1990. *Conservation Reserve Program "Logo Package."*

the almost 70 million acres of cropland seriously affected by wind erosion.[45] Nearly two million additional acres could be required to protect just 10 percent of other croplands, homes, and farms and to serve as living fences along rural roads. By one estimate, if this many trees were planted, almost 100 million tons of carbon could be stored over time directly in biomass and the emission of almost 165 million tons of fossil fuel carbon prevented over 50 years.[46] It is worth noting, however, that large numbers of windbreaks have been cut down in recent decades to make way for larger fields and irrigation systems.[47]

Overall, the technical potential for land conversion runs into the tens and possibly even hundreds of millions of acres. If 50 million acres were converted, 50 to 125 million tons of carbon could be stored per year, up to a total of 2.5 to 6.3 billion tons over a 50-year growing cycle. Converting 100 million acres of land could lead to the storage of 100 to 250 million tons of carbon per year—a significant fraction of current U.S. fossil fuel emissions, some 1.6 billion tons. These figures dwarf the estimated potential of existing forestry-incentive programs to encourage additional tree planting. Only an estimated 300,000 additional acres could be planted annually under the ACP and the FIP. And according to USDA's Economic Research Service (ERS), paying all tree-establishment costs and extending CRP contracts for 10 years from the date of tree planting could encourage the afforestation of just 0.5 to 1.2 million acres of former cropland already enrolled in the CRP, a small fraction of the almost 40 million acres enrolled in the program.[48]

The Costs of Converting Private Lands. The physical potential for reforesting pasture and agricultural lands is huge, but how much would it cost? One consideration is the various incentives that are encouraging the agricultural use of lands that might otherwise be forested or reforested. Almost 70 percent of U.S. cropland is currently enrolled in one federal commodity program or another.[49] Overall, the federal government spends more than $10 billion

each year on farm programs, some proportion of which fosters the use of marginal and environmentally fragile lands.[50] Changing the terms or scope of these programs might encourage considerable tree planting while actually saving taxpayer money. In addition, most schemes proposed for reforesting private lands have focused on direct government payments to landowners. Many other approaches, including higher taxes on fertilizers and other inputs, might have different costs and impacts.

How much would it cost to impose a new layer of policies to encourage the conversion of certain categories of agricultural land to tree cover? Past experience with such programs should provide rough ideas. Some 4.1 million acres have been converted to tree cover through the ACP, FIP, and CRP over the last decade. If tomorrow's reforestation efforts take the same basic approach on the same land base, per-acre costs will no doubt continue to rise. Loosening eligibility criteria would lower program costs, though secondary environmental benefits would be lessened if the trees are planted on hardy, resilient lands. The choice between offering only cost-sharing payments (as the ACP and the FIP do) or making rental payments as well (as the CRP does) also significantly affects projected program costs. The average federal share of tree-planting costs on the 1.6 million acres enrolled under the CRP as of 1988, for example, was a one-time $35 per acre, or $56 million. In contrast, rental costs for those acres will average $45 to $50 per acre per year for 10 years, for a total cost of $70 to $80 million per year.[51]

A variety of estimates have been made of the cost of using government payments to expand reforestation. USDA's ERS believes that expanding the FIP and ACP by 300,000 acres per year would cost some $25 to $35 million annually if the government pays all planting costs.[52] Another ERS analysis suggests that $123 to $239 million would be needed to convert 0.5 to 1.2 million acres of land already in the CRP to tree cover if the government pays all tree-planting costs and extends CRP contracts for an additional 10 years.[53] Depending on how fast the newly planted trees grow, the annualized cost per ton of carbon stored is likely to range from $5 to $20 under these scenarios.

The ERS has considered even larger-scale efforts. A new Forest Reserve Program for converting 10 million acres would cost an estimated $2.2 billion, and $5.3 billion would be needed for 20 million acres—roughly $10 to $30 per ton of carbon stored.[54] More recently, economists Robert Moulton and Ken Richards of the USFS argued that up to 250 million acres of U.S. cropland and pasture could be forested to store more than 700 million tons of carbon per year at a top marginal cost per ton of carbon of only $43 or so and a surprisingly low average cost of only $24 per ton.[55] These first-order estimates do not, however, take into account the major timber and agricultural market impacts of a program of this scale.

The secondary market impacts of large-scale tree planting, including depressing wood prices, are likely to prove quite important to the long-term carbon benefit of such programs.

The secondary market impacts of large-scale tree planting, including depressing wood prices, are likely to prove quite important to the long-term carbon benefit of such programs. Unless the timber grown in large-scale projects is permanently kept out of the market, which would in itself raise the price demanded by farmers to convert their lands, such afforestation projects would probably push forestry investment down elsewhere in the country.[56] If one-half the carbon accumulation on the converted lands were negated by forgone tree growth on other lands in the United States or abroad, the real cost per incremental ton of carbon stored would double.

Despite the confusion engendered by varying estimates and the assumptions behind them, in principle it should be quite cheap to convert some 20 million acres to tree cover because landowners would make more money from planting timber on these lands than they now do grazing cattle on them. Additional tens of millions of acres could be included in a Forestry Reserve Program for only $10 to $30 per ton of carbon stored if the appropriate sites can be cost-effectively included. True costs might be even lower if the future financial returns from tree planting are positive and if savings result from reducing federal commodity price payments and curbing soil erosion. On the other hand, uncertainties about likely carbon accumulation rates, as well as the schemes' second- and third-order effects on timber and food markets, make it difficult to project reliably the net carbon storage or costs for these larger-scale conversions. In principle, however, the opportunities for private land conversions, ranging from windbreaks to plantations, are large.

The Pros and Cons of Converting Private Lands to Forest Cover. Besides carbon sequestration itself, the benefits of converting cropland or pasture vary considerably with the circumstances and the scale of the conversion. Reforesting highly erodible croplands and those that are major nonpoint sources of pollution from agricultural runoff, or planting trees in windbreaks and along riparian habitats, would be more environmentally beneficial than planting ecologically hardy lands—one reason to encourage additional tree planting on fragile lands. The land base originally eligible for inclusion in the CRP, for example, accounted for 60 percent of the total sheet, rill, and wind erosion from U.S. croplands.[57] Unfortunately, some of the more promising planting strategies have not been accorded much attention even under the CRP: today, windbreaks constitute only about 6,800 acres of more than 2 million converted to tree cover under the program.[58]

An additional advantage of land conversions like these is that the return on public investment

29

has tended to be long lasting without continuing governmental oversight, partially because the farmers involved have realized net economic gains beyond the government payments received.[59] According to one assessment, 86 percent of the acreage planted under the Soil Bank program is still in tree cover after 15 to 20 years.[60] On forested ACP acreage, over 90 percent of the trees planted are still there after 10 years.[61] Acreage forested under the CRP programs is likely to be even more permanent considering the constraints placed on returning highly erodible land to agricultural production under the 1985 Farm Act. Nevertheless, if long-term drought in the U.S. wheat belt persists, or if the United States fails to produce enough wheat or other staples to feed even itself, returning withdrawn lands to agricultural production might once again become politically irresistible. And if enough trees are planted to affect timber prices and financial returns to landowners significantly, landowners' incentive to keep converted lands forested once government payments expire may evaporate. Under such circumstances, longer-term contracts or easements could be required to preserve the carbon-storage benefits associated with large-scale tree planting.

An inherent risk with any large-scale rural tree-planting effort is its susceptibility to natural events.

An inherent risk with any large-scale rural tree-planting effort is its susceptibility to natural events. Droughts, hurricanes, disease outbreaks, or wildfires could all acutely affect how much carbon is stored on converted lands over a given time. As noted, for example, hurricane Hugo downed millions of acres of timber overnight. Although experience suggests that physical threats of this sort tend to damage seriously only a relatively small proportion of

forested land each year,[62] global warming could make such events more frequent and severe.

Another potential disadvantage of adding new incentives to influence land-conversion and land-management practices is the potential problem of overlapping incentives. In such a case, landowners might receive more money by enrolling in a new program without planting more trees than they would have under an existing program. However, already existing incentives that could in principle overlap in this way—reforestation tax credits and cost sharing, for instance—have thus far had an additive effect on encouraging reforestation.[63]

Large-scale land conversions could also take a toll on the health of rural economies and the industries they supply while also depressing timber markets. Covering 200 million acres of crop and pasture lands with highly productive plantations, for example, could almost double total commercial wood production from U.S. forests and produce more than 200 million tons of carbon in the form of merchantable timber per year. It could also produce tens of millions of tons of nonmerchantable wood that could be used to produce energy, just as wood wastes from current timber harvesting supply a large fraction of the processing industry's energy needs. Large-scale land conversions of even a fraction of this magnitude, however, would inevitably be offset to some extent by current timberland owners deciding not to reforest their own lands because of a perception that future timber prices would fall. According to one analysis of world timber supplies, planting 37 million additional acres of plantations in developing countries would yield 10 billion cubic feet of timber per year but would increase global timber production by only 2 billion cubic feet.[64] Under this scenario, as much as 80 percent of the mathematically calculated carbon benefit of the new plantations could be negated by reduced investment in forest regeneration elsewhere in the world.[65]

Given the vast areas of private land that potentially could be converted to tree cover, the barriers to large-scale forestry for global warming mitigation are more economic and social than physical. Some farmers and other landowners will no doubt resist becoming tree farmers. And the amount taxpayers would be willing to spend to induce landowners to convert millions of acres of economically valuable land to tree cover, or which agricultural policies that inflate the area of land in agricultural use could be eliminated, is unknown.

Given the vast areas of private land that potentially could be converted to tree cover, the barriers to large-scale forestry for global warming mitigation are more economic and social than physical.

Overall, tree planting on private lands could in principle sequester or offset more than 900 million tons of carbon per year. Even an aggressive land-conversion program, however, is unlikely to result in the net storage of more than a fraction of this technical potential. Indeed, the conversion of 30 to 50 million acres of land, resulting in the annual storage of an additional 30 to 125 million tons of carbon per year—an amount roughly the equivalent of two to eight percent of current fossil fuel emissions—could be seen as a significant success. Even at the upper end of this relatively modest range, commercial timber production might increase by as much as 7 billion cubic feet per year, a significant fraction of today's 18 billion cubic feet. This change could well cause market feedbacks such as higher land and lower timber prices that would partially negate the carbon-storage benefits.

3. Increasing Tree Cover on Nonforested Public Lands

Hundreds of millions of acres of nonforested land in the United States are managed by government agencies. The Bureau of Land Management (BLM) alone controls 270 million acres of public lands, mostly pasture and rangelands. The Department of Defense (DOD) holds title to an additional 26 million acres. No doubt some public lands, including active and closed military bases, lands administered by the BLM, and highway and pipeline rights of way, would be hospitable to additional tree planting or other types of vegetation.

The Potential Magnitude of Afforestation on Public Lands. Relatively little information exists on the potential for increasing forest cover on public lands, and the recently announced America the Beautiful campaign makes no provision for such planting. Currently, less than 30 million acres of BLM lands are forested, and less than 10 million are considered productive timberland.[66] Some 11 to 12 million acres of DOD lands are forested.[67] Preliminary Environmental Protection Agency (EPA) analysis suggests that improving forest management or planting trees on one to three million acres of BLM lands would result in the storage of an additional 1.5 to 4 million tons of carbon per year.[68] In the same analysis, it was assumed that five percent of the Department of Defense's 26 million acres could be replanted or fully stocked, with a carbon benefit of one to three million tons of carbon per year. EPA has also estimated that 20 percent of the land area currently dedicated to rural highways (other than the roads themselves) in the northeast, north-central, and southern parts of the United States—some 4.5 million acres in all—might be suitable for reforestation.[69]

Relatively little information exists on the potential for increasing forest cover on public lands.

The Costs of Public Land Conversions. Reliable cost figures for each ton of carbon stored

31

through land conversion on public lands are not available. Planting and managing trees on rights-of-way would probably entail high labor and energy costs. Although land purchase or rental payments would not be required to encourage tree planting on government lands, tree-establishment costs are likely to be higher than on private lands, many of which are more suitable for trees.

The Pros and Cons of Public Land Conversions. The primary advantage of converting public lands to forest or other tree cover is that no land purchase or rental payments would be required. But many unforested federal lands simply cannot support forest. Many of BLM's rangeland acres and many DOD acres, for instance, are too arid for tree planting. Some may also be degraded by past misuse or by military exercises. The BLM claims that no more than an additional 50,000 acres of its lands could be forested. The Forest Service also argues that it has no unforested acreage to plant, though in the past the Forest Service has administratively withdrawn logged-over lands from its reforestation backlog rather than replant them.[70] Physical limits to tree planting probably are not the primary barrier on federal lands, however; it is that government agencies do not currently see carbon storage as part of their missions.

Probably the least contentious way to increase carbon storage on unforested public lands at least marginally is to increase the number of trees found on rights of way, particularly along rural and interstate highways. Department of Transportation analysts, however, suggest that much less land is available than might be supposed.[71] Once the land dedicated to driving lanes, shoulders, median strips, and clear zones is discounted and certain areas are eliminated for safety reasons, the gross total available for tree planting along U.S. rural interstate and primary highways may not exceed 1.1 million acres. Of this, some 30 percent is believed unsuitable for tree growing, and at least some significant fraction of the remaining 710,000 acres is already forested.[72]

For safety, political, and economic reasons, proposals to harvest such timber commercially to defray costs also seem highly impractical.

Overall, the carbon-storage opportunities on unforested public lands appear quite modest— only enough to sequester perhaps three to five million tons of carbon per year. This policy option, however, remains considerably less well investigated than the conversion of privately held lands, and further research could well reveal further potential.

4. Stocking and Managing Already Forested Lands More Intensively

Yields on many of the 481 million acres of U.S. timberlands are well below physical potentials. To maintain or increase carbon-storage capacities, options range from forgoing the harvesting of some timberlands to clearing and replanting others.

Selectively forgoing timber harvesting to preserve existing carbon reservoirs can prolong the storage of carbon that would otherwise be released in the short term, particularly where the ratio of standing biomass to annual growth is unusually high—as it is, for example, in the old-growth forests of the Pacific Northwest and Alaska. Old-growth forests can contain hundreds of tons of carbon per acre in trees and soils.[73] For every acre of these lands left uncut, more than 220 tons of carbon can be kept out of the atmosphere in the short term. *(See Box 3.)*

Nationally, however, forest ecosystems store only about 80 tons of carbon per acre. Leaving lands with these lower carbon densities uncut for carbon storage is generally much less compelling, both because replanting the land can often restore the preexisting carbon stocks within a matter of decades and because forgoing the harvest could increase significantly the use of plastics and other alternative construction materials that require fossil fuel energy to manufacture. This statement does not suggest that restricting the harvest of forest other than old-growth is never justified in the name of carbon storage. Below-cost timber sales, for example, provide an interesting example.[74] Although how such sales affect the long-term carbon cycle is likely to vary by forest,[75] many below-cost sales occur in relatively unproductive forests that will take decades to regenerate, if they successfully regenerate at all.[76] In such forests, soils are likely to release a greater proportion of their carbon to the atmosphere in the interim than is the case with more productive land. In addition, cutting these forests could be expected to release relatively more CO_2 through biomass oxidation than would occur in more productive forests (where a greater proportion of total biomass would likely be tied up in merchantable timber and where the collection and utilization of wood wastes might be more economically justified).

How below-cost timber sales affect the long-term carbon cycle is likely to vary by forest, since many below-cost sales occur in relatively unproductive forests that will take decades to regenerate, if they successfully regenerate at all.

Making timberlands more productive is another approach to enhancing carbon storage both in the trees themselves and in wood products. Currently, the great majority of U.S. timberlands regenerate naturally once cut and are not intensively managed during the growth cycle. But they could be replanted, possibly with genetically improved seedlings, thinned during the growing cycle, or even fertilized. Timber stands can be intensively managed to protect against fire, insects, disease, and poor logging practices. Further reliance on the increasingly controversial even-aged stand management practices that have characterized

forest management in recent decades is not necessary to increase carbon storage.[77] What is needed instead is considerable site-specific research to determine how the goal of carbon storage can be integrated into existing multi-use forest management and to determine the capacity of U.S. timberland soils to sustain increased production.

The Potential Size of Carbon Benefit from Forest-Management Changes. Although the extent and even the definition of old-growth forest are in dispute, as much as four million acres of what is conventionally considered old-growth forest could be subject to harvest over the next several decades.[78] If a 50-year moratorium on harvesting were called to preserve standing carbon stocks, the net carbon benefit could total 20 million tons per year in the short term and more than 11 million tons per year over the long term—the equivalent of the CO_2 emissions of some 4,000 megawatts (MW) of coal-fired electricity.

Managing timberlands more intensively is a carbon-storage option that has received much more attention than that of putting some timberlands off limits altogether. Researchers have, for example, estimated that U.S. timberlands could theoretically produce more than twice as much merchantable timber as they do today.[79] *(See Figure 7.)* In most cases, the shift to more intensive management would be gradual and would begin with a new rotation cycle. Alternatively, particularly poorly stocked stands could be restocked without waiting for a future harvest. Estimates of how much timberland could practically be brought under intensive management within several decades vary from 45 to 150 million acres.[80] Such a program could be jump-started by reducing today's backlog of harvested acres waiting reforestation—about 1.2 million acres, nearly a three-year supply at current planting rates. Because federally owned timberlands constitute less than 20 percent of total U.S. timberlands, their likely contribution to strategies for increasing carbon storage in existing forests is modest compared to that of private lands.

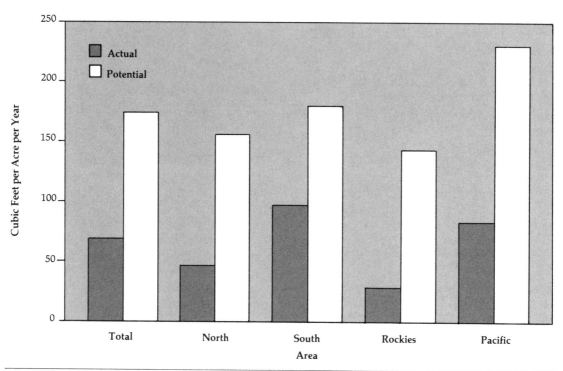

Figure 7. Growth Potential on United States Timberlands

Source: U.S. Department of Agriculture, Forest Service, *An Analysis of the Timber Situation in the United States: 1952–2030,* Forest Resource Report No. 23, (1982)

The degree to which increases in commercial timber production might simply supplant the growth of noncommercially useful trees or species, thus failing to add to net carbon accumulation, cannot accurately be estimated.

Opportunities for intensifying forest management are abundant. The Forest Service estimates that if only those management changes justified by current market economics were made, timber production on private timberlands would increase by almost four billion cubic feet per year—the equivalent of some 37 million tons of merchantable carbon, or an increase of some 30 percent over current commercial productivity on these lands. Because only three percent of the eight million private small-scale timberland owners in the United States have taken advantage of federal assistance to enhance timberland productivity, there is considerable room for improvement.[81] The degree to which increases in commercial timber production might simply supplant the growth of noncommercially useful trees or species, thus failing to add to net carbon accumulation, cannot accurately be estimated.

Timelines are quite important in assessing the practical potential of intensifying forest management expressly to store carbon. If all harvested acres for the next 25 years were to be replanted under intensive management, the acreage under such management would increase by more than 75 million acres. If net carbon storage on these acres were to increase by 0.5 to 1 ton per acre per year,[82] total carbon storage could increase by 35 to 75 million tons per year for several decades. Because intensive management will not be appropriate for all acres and because the time horizon for this analysis does not extend more than several decades into the future, it appears realistic to estimate the carbon benefit that might accrue from this policy option in the near- to midterm to be in the 15 to 20 million tons-per-year range. Roughly speaking, this 15 to 20 million tons could offset one percent of current carbon emissions from U.S. fossil fuel combustion.

The Costs of Intensifying Forest Management. Like the policy options already discussed, the costs of managing forests to increase overall carbon storage are hard to specify. Perhaps most difficult to gauge are the true costs of prohibiting logging on additional acres of federal (or even private) old-growth timberlands. One way to calculate the cost of this option—yielding a figure of $100 to $150 per ton of carbon "locked up"—is to base it on the current market value of an acre of high-quality old-growth timber.[83] It is this calculation that will convince many observers that the price is too high, especially when mill towns and logging communities are already suffering layoffs and shutdowns in the Pacific Northwest. On the other hand, other variables, including mechanization in the timber industry and the loss of processing jobs to the Japanese (who import raw logs from the United States and process them in Japan)—rather than preservation of old-growth forests for carbon storage or any other purpose—are also behind the economic crisis in timber-dependent communities in the Pacific Northwest.[84] In any case, the outcome of current economic and employment trends is likely to be a restructuring of the timber

harvesting and processing sectors in the relatively near future. Putting old-growth lands off limits may only accelerate the inevitable. In addition, biological and recreational values alone might economically justify logging bans in many old-growth forests. From this perspective, prohibiting logging in old-growth forests could offer a cheap approach to storing carbon.

After accounting for biological diversity, prohibiting logging in old-growth forests could offer a cheap approach to storing carbon.

Another cost consideration is the money required to intensify forest management. Basically, the more intensively timberland is managed, the more it costs. Enticing private landowners to intensify management on their lands has in the past involved governmental cost-sharing and, in some cases, rental payments as well. The costs of forest-stand improvement can run to $450 per acre, although a more common range is $100 to $150.[85] Examples of management practices include site preparation and artificial regeneration, use of genetically improved seedlings, stand fertilization, precommercial thinning, and timber salvaging. If timber productivity were to increase by 0.5 to 1 ton of carbon per acre per year as a result, the cost per ton of additional carbon stored could be $10 to $100.[86] Fortuitously, intensifying forest management appears justified on many acres by existing market incentives, and it could presumably be achieved at a relatively low cost.

The Pros and Cons of Changing Forest Management Practices. Reducing the harvest of certain forest types serves multiple goals in areas where pressure to constrain timber harvests for biological, aesthetic, or recreational reasons is already mounting. Increasing timber growth on other forested lands does not require changing land uses, and this approach conforms well to the mission of the USFS and other agencies likely to be involved. Expanding the extension services and the capabilities needed to encourage better management of private timberlands is also relatively easy.

Global warming itself may frustrate efforts to preserve large stores of carbon in old-growth forest.

From a carbon-storage perspective, however, legitimate questions can be raised about some of these options. First, just as the calculated carbon benefits of tree-planting on nonforested lands could be lost if forestry investment dropped elsewhere in the United States or abroad, the calculated benefits of forgoing harvesting could be dissipated if the harvest of old-growth forests increased elsewhere in the world as a result. Second, global warming itself may frustrate efforts to preserve large stores of carbon in old-growth forest. Indeed, forestry scientists expect to use old-growth forests as global warming laboratories precisely because they are so susceptible to damage from climate change.[87]

Proposals to restock existing stands to promote timber growth must be evaluated with particular care. If these stands already contain considerable biomass, albeit noncommercial species or trees, restocking may be a wash from a carbon-sequestration standpoint or may even release carbon on net. Indeed, recent analysis suggests that restocking many poorly stocked U.S. forests might be commercially worthwhile but would yield little, if any, carbon benefit.[88]

Forgoing harvesting and intensifying management of already forested lands have other drawbacks as well. Hopes of mitigating global warming are unlikely to move local logging communities to accept reduced timber harvests. Imposing a carbon-storage mandate onto forest management agencies, whose performance has been measured previously largely in terms of board feet harvested, will also involve changing entrenched and sometimes complex decision-making processes. In addition, intensifying timber management to enhance growth may require environmental tradeoffs, and issues such as soil capabilities, preservation of wildlife habitat, and fulfillment of other multiple-use objectives have to be balanced carefully. Intensifying forest management could make timber management less sustainable than it is now or work against other policy objectives; balancing these interests may prove just as contentious as forgoing the harvesting of the much smaller numbers of acres for which this option would be appropriate.

Recent analysis suggests that restocking many poorly stocked U.S. forests might be commercially worthwhile but would yield little, if any, carbon benefit.

---

5. Substituting Wood for Other Products

Although relatively little is known about the life span and ultimate fates of various wood products, such products do constitute an artificial carbon sink. Storing harvested carbon in products can postpone the carbon's return to the atmosphere through oxidation, which happens relatively quickly when a dead tree is left to decompose in the forest. Increasing the use of wood products can also defray fossil fuel emissions to the extent that manufacturing nonwood substitutes like plastics and concrete requires comparatively more fossil fuel.

Although the average useful life of wood products can be reasonably well estimated, the carbon storage of these products remains difficult to ascertain. It is known that a huge stock of worn-out wood products can now be found in U.S. landfills and that decomposition there is slow, but research on how much carbon is actually released in the use and disposal of wood products and the mechanisms through which it is released is just beginning.[89]

Increasing the use of wood products can also defray fossil fuel emissions to the extent that manufacturing nonwood substitutes like plastics and concrete requires comparatively more fossil fuel.

Carbon sinks in long-lasting products can be expanded by increasing timber harvest levels, the percentage of harvested biomass that goes into long-term product, or the life spans of the products. Currently, considerably less than one-half of all harvested wood is used directly in long-lived wood products.[90] More efficient harvesting, milling, manufacturing, and building practices could increase this percentage. Reusing wood materials and extending their life spans through the use of preservatives and other means could increase the time it takes for products to reoxidize to the atmosphere. Using recycled materials to produce larger proportions of paper and paperboard could also make more lumber available for use in long-lived products by reducing demand for pulp. A drop in demand for virgin fiber, however, could just as easily decrease overall forestry investment and long-term carbon storage.

The Pros and Cons of Increasing Timber Utilization in Long-Lasting Products. Too little is known about the opportunities for using wood products as carbon sinks to draw any firm conclusions about the option's potential, side effects, or cost-effectiveness. There is a practical limit to the amount of wood that can

accumulate in product stocks (including in landfills) before annual oxidation equals or exceeds the annual addition of fresh product, and researchers do not know how far from this point the United States is now. Further complicating matters, as wood products decompose, they can also produce methane, which is a much more damaging greenhouse gas than CO_2. Little information is available, however, on the rates and patterns of wood fiber decomposition in landfills, where the anaerobic production of methane is common. For this reason, any estimates of either the technical or practical potential for this option are premature.

6. Increasing the Commercial Use of Biomass Energy

Biomass energy has been used for millennia. Ever since firewood was first burned for heating and cooking, human societies have relied on the energy embodied in vegetation and other organic matter. Until the relatively recent mass exploitation of fossil fuels, biomass was the world's chief energy source. Even today, more than one-half the wood removed from U.S. forests is used as firewood, though the total energy produced from wood accounts for only four percent of energy use in the United States.

Although our relative reliance on biomass has decreased markedly since industrialization began, the technological options available for converting biomass to useful energy have multiplied.[91] Through techniques ranging from direct combustion to thermal gasification of feedstocks—be they annual grasses or whole trees—biomass can be converted into electricity, ethanol, methanol, biocrude gasoline, diesel fuel, or even synthetic natural gas. Given this versatility, biomass crops are not valuable primarily because they store carbon themselves but because they possess the potential to reduce the need for fossil fuels. The relatively high growth rates of biomass energy crops make them more suitable for energy generation than conventional tree cover in many cases, and their relatively short rotations make it easier to adapt them to changing environmental and climatic conditions.

Substituting biomass for fossil fuels does at least as much to mitigate global warming as would conserving the displaced quantity of fossil fuel.[92] And by raising the amount of carbon stored in the soils and vegetation of an average acre above what it is in conventional agricultural crop production, woody and perennial energy-cropping systems help expand natural carbon reservoirs and mitigate global warming even further.

Programs for the large-scale tapping of biomass energy, however, are not yet commercial for both economic and technological reasons. To help make biofuels more competitive with conventional fuels, researchers are striving to increase the production rates of new feedstock crops over a broad range of site types, reduce management and harvesting costs, and improve the quality of biomass so that it can be converted to fuel more efficiently. The U.S. Department of Energy's (DOE) work in energy crops emphasizes short-rotation woody crops (SRWC) and perennial herbaceous species—including switchgrass, with productivities of almost 2.5 tons per acre in a drought year with little fertilizer—and a tropical sugar cane hybrid that has yielded as much as six tons per acre per year.[93] Research on woody crops centers on poplar and sycamore species because they grow so rapidly and can regenerate from stumps and roots. To maximize biomass production, species such as poplars are planted at a density of 700 to 1,600 or more seedlings per acre, somewhat higher than in a conventional pulp or timber plantation, and harvested every four to eight years. Mean annual productivities of 2.8 to 3.5 tons of carbon per acre have been achieved with these trees in several locations, and a yield of 8.5 tons per acre per year was recorded in one field trial. DOE's target is to achieve five tons of carbon per acre per year over a wide variety of sites.

The Potential Magnitude of Future Biomass Energy Use.
The amount of land that could in principle be shifted over to biomass production and the amount of energy that could ultimately be produced are limited by economic and social factors, not physical constraints. Although more than 200 million acres of land not currently used for agricultural crop production appear suitable for short-rotation cropping, estimates of how much land could be practically committed to this use without cutting significantly into productive agricultural land range from 75 to around 100 million acres.[94] And if agricultural production continues to become more concentrated, more land would become potentially available for biomass energy production. Many of the determinants of land availability for converting cropland and pasture to conventional tree cover must also affect the availability of land for short-rotation wood or herbaceous energy-crop production because the available land base is essentially the same. One difference, however, is that over time energy crops would be more likely than conventional tree crops to compete for higher-quality cropland because they could offer farmers greater flexibility and faster returns on their investments.

Estimates of the technical potential of biomass fuels to offset fossil fuel emissions vary widely, largely because so many variables are involved. Biomass growth rates, land availability, energy inputs into the growing cycle, the energy requirements of converting biomass to alternative fuel types—all can significantly affect how much fossil fuel carbon is actually displaced when carbon is drawn out of the atmosphere by biomass energy crops. The production of ethanol from corn, for example, requires almost 50 percent more energy (excluding solar energy inputs) than is contained in the liquid fuel produced.[95] From a global warming perspective, ethanol from corn is thus undesirable.

A wiser tack is using less-energy-intensive crops and more efficient energy-conversion technologies. They can help reduce the amount of fossil fuel energy required to produce the crop in the first place as well as the proportion of energy stored in the crop that is lost during the production of the final biofuel product. In the case of synthetic natural gas (SNG) production from herbaceous species, only 30 percent of the energy ultimately produced can actually count toward fossil fuel displacement.[96] In ethanol production from woody energy crops, this figure rises to 75 percent, and when such crops are directly converted to electricity, the figure can rise to 90 percent.[97]

A 1975 analysis of SNG production from croplands estimated that converting some 75 to 100 million acres would yield 7 to 11 trillion cubic feet of synthetic natural gas—35 to 60 percent of current natural gas consumption in the United States.[98] More recently, the DOE has estimated biofuels' potential at about 17 quads per year, roughly 20 percent of total U.S. energy consumption. Nearly one-half this potential under the DOE scenario, however, would come from conventional wood and forest wastes.[99] According to another estimate, agricultural and municipal wastes, bioenergy crops, and other forms of biomass could eventually displace up to one-third of all U.S. fossil fuel use, including 30 to 90 percent of the gasoline and diesel fuel used for transportation.[100]

With current technology, converting 35 million acres of land to SRWC (with a productivity of 2.7 tons of carbon per acre per year) would on net displace consumption of 67 million tons of fossil fuel carbon if the biomass were directly transformed into electricity.

The most comprehensive analysis of the fossil fuel displacement potential of woody short-rotation crops has been performed by Lynn Wright and her colleagues at Oak Ridge National Laboratory, a team that has also developed scenarios of alternative biofuels futures.[101] With current technology, they estimate, converting 35 million acres of land to SRWC (with a productivity of 2.7 tons of carbon per acre per year) would on net displace consumption of 67 million tons of fossil fuel carbon if the biomass were directly transformed into electricity (34 million tons if converted to ethanol). These figures work out to a fossil fuel displacement rate (taking into account energy inputs and conversion losses) of some 1.9 tons of carbon per acre per year for electricity production and almost 1 ton of carbon per acre per year for ethanol production. If assumptions regarding advanced conversion technologies are used and 5 tons of carbon can be fixed per acre per year, net fossil fuel displacement rates per acre increase by approximately 125 percent in each case. Considering a larger land base, 103 million acres, the Oak Ridge researchers conclude that (with advanced technologies and high growth rates) the emission of some 470 million tons of carbon from fossil fuels could be offset if the biomass is used to produce electricity and 272 million tons if converted to ethanol as a transportation fuel. This amount is the equivalent of more than 80 percent of the total CO_2 released by electric power production in the United States. The same amount of biomass converted to ethanol using advanced conversion technologies could satisfy 79 percent of current gasoline consumption—indefinitely if use rates remained stable. If the more than 250 million acres of private land discussed previously as potential candidates for conversion to tree cover were all considered eligible for conversion to biomass energy production, the carbon benefit figures could rise considerably higher.

The Costs of Increasing Biomass Energy Use. Whether sophisticated biofuels will be used more in the future depends substantially on comparative energy costs. Although biofuels' production costs have dropped significantly over the last two decades, it will probably be years before most biofuels can compete with fossil fuels in the absence of further policy encouragement or sharp fossil fuel price hikes.

Gasoline at $1 and $1.50 per gallon is the equivalent of $8 and $12 per million Btu (MMBtu), respectively. According to DOE estimates, the cost of converting wood to a liquid transportation fuel (biogasoline) could drop from $2 to $3 per gallon or $16 to $20 per MMBtu today to around $0.85 per gallon or $7 per MMBtu by 2007. With further research, the cost of wood-to-ethanol production is also expected to drop, from roughly $25 per MMBtu today to under $10 per MMBtu in the next decade.[102]

More sophisticated biofuels are even more expensive, though prices are dropping fast. The estimated costs of a microalgae-conversion facility have fallen from $149 per MMBtu in 1982 to $58 in 1987, and they are expected to drop to $13 by 2010.[103] The cost of oilseed diesel fuel is projected to drop from $18 per MMBtu in 1987 to $8 in 1995, the equivalent of dropping from $2.75 to $1.50 per gallon.[104]

The direct conversion of short-rotation biomass to electricity presents the greatest short-term opportunity for substituting biomass energy for fossil fuels on a significant scale, though the relative costs of biomass production remain high. With yields of 2.5 to 3.5 tons of carbon as biomass per acre on a good site, the cost per dry ton of wood chips delivered to a powerplant would range from $35 to $50, the equivalent of $2 to $3 per MMBtu. Whole-tree handling (rather than chipping prior to combustion) might reduce feedstock costs from good sites to $1.50 to $2.00 per MMBtu.[105] Larger-scale woody biomass production on a broader range of site types, however, would raise the average delivered cost to approximately $4 per MMBtu. For perspective, coal is currently being delivered to powerplants at an average national cost of $1.25 per MMBtu, while natural gas is being delivered for about $2 per MMBtu.[106]

Subsidizing large-scale biomass energy production or, alternatively, taxing fossil fuels could make biomass competitive with fossil fuel feedstocks for electricity generation. At current market prices, the cost differential between large-scale biomass production and coal is $60 to $90 per ton of carbon, roughly the equivalent of $1.75 to $2.75 per MMBtu. This sum does not reflect the cost of producing the biomass fuel but, rather, the net differential required to allow it to compete with coal—roughly the equivalent of a tax of two to three cents per Kwh of electricity consumed. Over the next decade, this differential should continue to fall, thus making large-scale biomass production more economically viable. Already, biomass production is nearly competitive on a few million acres of good sites where the required differential is much smaller, perhaps as low as $10 to $15 per ton of carbon.

Such a wide range of technologies can be used to convert biomass of various types into fuel that broad generalizations about the pros and cons of increasing commercial biomass utilization cannot be made.

The Pros and Cons of Increasing Biomass Energy Use. Such a wide range of technologies can be used to convert biomass of various types into fuel that broad generalizations about the pros and cons of increasing commercial biomass utilization cannot be made. Research on energy crops is too young to determine these crops' long-term ecological and land-use impacts, though available evidence suggests that energy cropping is more environmentally benign than conventional cropping systems. Perennial energy crops require less tillage (thus reducing erosion) and smaller amounts of chemical additives (thus reducing the risk of contaminating the soil, water, and farm workers) than most agricultural crops. On the other hand, biomass production is currently considerably more occupationally dangerous per unit of energy produced than is the production of fossil fuels.[107]

Whatever the advantages of increasing biomass use, any large-scale expansion is still some years away. The practical potential of biomass fuels over the next decade or two is likely to be considerably smaller than the long-term potential calculated by Wright and colleagues, particularly for applications other than direct electricity production.[108] Although direct biomass-to-energy combustion techniques (primarily using chips and wood wastes) are in limited use, more research and development is needed on these and on more complex energy-conversion technologies. It will probably be quite a few years before large-scale liquid-fuel conversion plants are designed and built, but making direct-combustion technologies more efficient could yield benefits almost immediately.[109] One particularly promising concept is the whole-tree burner.[110]

Depending on how comprehensive our energy policy is and how big the subsidy or tax used to encourage biomass competitiveness, biofuels could displace annually the consumption of 20 to 150 million tons of carbon from fossil fuel combustion within 15 to 30 years.

Although the theoretical potential of biofuels to displace carbon emissions from fossil fuel combustion is likely to exceed one billion tons per year, the expectation of large increases in bioenergy production in the near term is unrealistic. It may be possible, however, to achieve 10 to 20 quads (or 12 to 25 percent of current U.S. energy consumption) within 30 to 40 years. Depending on how comprehensive our energy policy is and how big the subsidy or tax used to encourage biomass competitiveness, biofuels could displace annually the consumption of 20 to 150 million tons of carbon

from fossil fuel combustion within 15 to 30 years. For perspective, more conventional urban and rural forestry policies would also take about this long to deliver significant benefits. In fact, bioenergy production could begin to make at least a limited contribution to reducing U.S. CO_2 emissions before conventional trees would be large enough to begin to offset fossil fuel CO_2 emissions significantly. It may well be most cost-effective for now to develop energy-cropping programs only on the best sites and to grow trees conventionally on less-productive lands for timber production, some energy production, or purely carbon storage.

7. Improving Soil-Management Practices

Globally, two to three times more carbon is stored in soils than in above-ground biomass. *(See Box 1.)* Unfortunately, many land-use practices release soil carbon instead of retaining or augmenting it. Clearing and farming forested lands generally reduces soil carbon by up to 40 percent, and conventional monoculture tillage and harvesting practices work against the restoration of the lost soil carbon. Furthermore, conventional agricultural practices often allow or even invite wind or water erosion of fertile topsoil.

Changes in agricultural practice should make it possible to prevent continuing losses of soil carbon and to store additional carbon in soils by gradually rebuilding depleted soils. Rapidly reforesting logged lands, reforesting agricultural or pasture lands, and relying on resource-conserving tillage and crop rotations to reduce the need for such agricultural inputs as fertilizers and pesticides would all help to meet this goal.

The Potential for Storing Additional Carbon in Soils. More than 5 billion tons of topsoil were lost from farm and pasture acreage annually in the United States in the early 1980s—the equivalent of some 200 million tons of carbon. This figure is 10 to 50 times higher than the amount of soil formed during the same period.[111] Exactly how much of this carbon ultimately oxidizes to the atmosphere is not known, but the potential for slowing this loss or for increasing the amount of carbon stored in remaining soils appears significant. Indeed, a modest beginning has already been made through such programs as the CRP and other provisions of the 1985 Farm Act. Researchers can only speculate on how much more can be accomplished through better land management, integrated pest management, enhanced nutrient conservation, and soil and water conservation.

More than 5 billion tons of topsoil were lost from farm and pasture acreage annually in the United States in the early 1980s—the equivalent of some 200 million tons of carbon.

With modern sustainable agricultural practices, for example, it might be possible to build up organic matter in the soil by a small amount per year—say, 0.25 tons of carbon per acre.[112] On 300 million acres of agricultural lands, that increase could reach 75 million tons annually within the United States, in addition to whatever carbon oxidation is avoided by slowing erosion. Preventing wind and water erosion on these and other acres might double the net carbon benefit.

The Costs of Increasing Carbon Levels in Soils. It is estimated that soil erosion costs farmers anywhere from $1 billion to $18 billion per year.[113] Reducing erosion and otherwise enhancing the carbon content of soils could boost harvests, slow the siltation of water reservoirs, and reduce the need for fertilizer.[114] If the true social costs of erosion are near the upper end of this dollar range, major programs to stabilize and increase soil carbon levels are already economically justified. Recent WRI research suggests, for example, that when the long-term productivity changes owing to the degradation or improvement of soil structure are considered, chemical-free farming systems based on soil-conserving practices and crop rotations become more profitable than conventional corn monocultures or corn-soybean rotations that rely heavily on chemical inputs. If the off-site benefits—reduced soil erosion and reduced chemical runoff, among them—are also considered, the relative advantage of this approach is even greater.[115] On a cost-per-ton-of-carbon basis, the societal costs of this option likely range from negative to modest.

The Pros and Cons of Increasing Carbon Levels in Soils. Programs to enhance soil conservation traditionally have been adopted slowly and piecemeal, partly because federal incentives often discourage such practices. Significantly changing these incentives is almost certain to prove a political battle. An alternative would be new government programs that independently would foster soil conservation, but the potential benefits and costs of such programs remain to be determined. With these political and bureaucratic hurdles, it seems unrealistic to expect to increase biotic carbon storage by more than 10 to 25 million tons of carbon per year through this mechanism in the near term.

IV. Conclusions

In theory, the biotic policy options discussed here collectively could absorb huge quantities of carbon annually, thereby displacing or offsetting a large fraction of U.S. fossil fuel emissions. Conventional tree planting and forest management are by themselves estimated capable of increasing biotic carbon uptake in the United States by more than 800 million tons per year. *(See Figures 8 and 9.)* Yet, technical potential is often quite different from practical reality. Indeed, the average commercial productivity of U.S. timberlands is just 0.6 tons of carbon per acre per year, compared to a technical potential of more than 3 tons of carbon per acre per year on good temperate soils.

In theory, the biotic policy options discussed here collectively could absorb huge quantities of carbon annually, thereby displacing or offsetting a large fraction of U.S. fossil fuel emissions.

Projecting the practical is far harder than projecting the technically possible. *(See Table 2.)* Over the long term, the United States is likely to possess the land resources and technology needed to pursue any given policy option to the absolute limit of its potential, if in fact that were perceived as desirable. What that limit may be remains uncertain in most cases, however, because of both physical and economic uncertainties. Nor is it likely that the United States would ever want to pursue any given option to the full extent of its theoretical potential; global warming mitigation is not the only operative policy goal. And none of the policy options discussed in Table 2 could be implemented in full instantaneously or even over the short term.

Projecting the practical is far harder than projecting the technically possible.

Arriving at what may be practically achievable by any given date, 2030 in the case of Table 2, therefore involves a significant element of subjectivity. The economics of various biotic policy options are not well understood and can change rapidly. Each policy option will face a rising marginal cost curve as more and more of the option's potential is used up. The perceived marginal cost curve for any individual option is dependent on whether the marginal costs reflected are private or societal, short term or long term. In addition, individual suboptions (e.g., biomass to electricity versus biomass to liquid fuel, harvesting planted trees versus leaving them standing indefinitely) all have their own marginal cost curves. The long-term

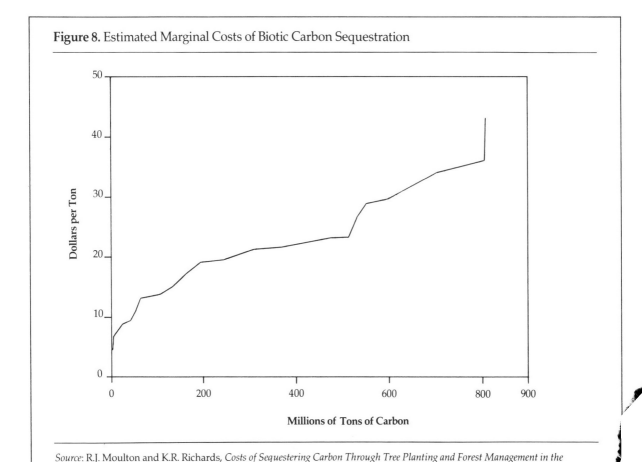

Figure 8. Estimated Marginal Costs of Biotic Carbon Sequestration

Source: R.J. Moulton and K.R. Richards, *Costs of Sequestering Carbon Through Tree Planting and Forest Management in the United States*, U.S. Department of Agriculture, Forest Service, (1990)

economic and environmental feedbacks of policy options have not yet been analyzed, and the existence of global markets for commodities such as timber will complicate such analysis. Market feedbacks from the large-scale implementation of biotic policy options anywhere in the world pose a potentially serious threat to the long-term effectiveness of carbon sequestration policies implemented unilaterally in the United States or any other country. Although energy policies pose some of the same problems—that is, any significant policy-inspired fall in global energy demand probably would be partially offset by falling prices and resulting upward pressure on consumption—these undesired market feedback effects are much easier to control in the energy sector; imposing

an energy tax that is neither too high nor too low can prevent the feedback altogether. In the forestry sector, however, imposing a guaranteed price floor for timber (the functional equivalent of the energy tax in this case) poses far more vexing problems.

Reliable answers to these questions will elude observers and experts alike until carbon models incorporating the entire U.S. carbon cycle as well as international timber markets are built and validated—a fairly distant prospect considering how poorly many aspects of the U.S. biotic resource base are understood.

Research problems notwithstanding, biotic policy options do have a role to play as part of

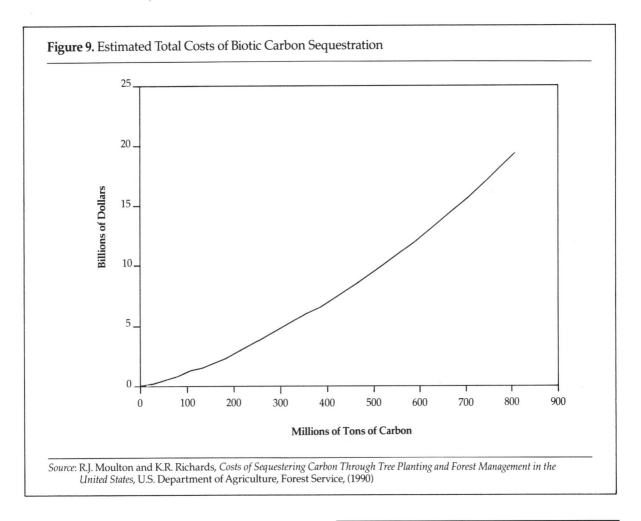

Figure 9. Estimated Total Costs of Biotic Carbon Sequestration

Source: R.J. Moulton and K.R. Richards, *Costs of Sequestering Carbon Through Tree Planting and Forest Management in the United States*, U.S. Department of Agriculture, Forest Service, (1990)

a comprehensive U.S. strategy to slow the buildup of CO_2 in the atmosphere. Although estimates of the size of this role must be treated cautiously, several important policy conclusions are not premature.

1. **Although biotic options offer considerable opportunities for sequestering or displacing carbon, do not expect their implementation in the United States to reduce net CO_2 emissions quickly or cheaply.**

Although implementation of any of the biotic policy options discussed here should reduce net U.S. CO_2 emissions from what would otherwise be the case, most of the carbon benefits would take so many years to realize that it is unlikely that they could stabilize net

Most of the carbon benefits would take so many years to realize that it is unlikely that biotic options could stabilize U.S. CO_2 emissions, much less reduce them from today's levels.

U.S. CO_2 emissions, much less reduce them from today's levels. Indeed, after 15 years of relative stability, U.S. industrial CO_2 emissions are once again increasing. *(See Figure 10.)* If nothing is done to interrupt this trend, annual U.S. CO_2 emissions could rise to almost 2 billion

Table 2. U.S. Biotic Policy Opportunities and Costs

Option	Theoretical Carbon Benefit[a] (millions of tons/yr)	Practical Benefit to 2030 (millions of tons/yr)	($/ton)
Urban Forestry	15	3–5	$0–25
Private Land Conversions	400–900[b]	50–150	$0–50
Public Land Conversions	(not known)	(not known)	(not known)
Forestry Management	100–400	35–75	$0–100
Forgoing Old-Growth Harvests	10–20	5–10	$0–100
Biomass Energy Production	400–1,000[b]	20–150	$20–75
Increase Wood Use	(not known)	(not known)	(not known)
Soil Carbon Buildup	50–150	10–25	$0–10

a. The carbon benefit figure estimates carbon storage in biomass and in soils as well as fossil fuel carbon displaced through implementation of the biotic options. Figures represent an average annual carbon benefit and thus overstate the carbon benefits in the early years of a policy option's implementation.

b. These entries cannot be summed because they involve use of overlapping land bases.

tons within the next 20 years.[116] Moreover, as emissions of CO_2 from fossil fuel combustion continue to rise, the relative role that could be played by biotic policy options in stabilizing or reducing net carbon emissions will inevitably shrink owing to a finite land availability.

Biotic options as a class are unlikely to prove cheaper than alternative approaches.

From the standpoint of cost-effectiveness, implementing some biotic policy options in some places is likely to pay for itself. Indeed, if all the federal programs that create disincentives to carbon storage—ranging from some below-cost timber sales to commodity support programs—were abolished, a great deal of carbon storage would appear cost-effective from the outset. Once these obvious options have been exhausted, the cost-effectiveness of biotic policy options will have to be compared with those from other sectors. Given the wide range of costs associated with these options and the similarly wide range often discussed in the context of energy policy options to reduce CO_2 emissions,[117] biotic options as a class are unlikely to prove cheaper than alternative approaches.

As noted, the geographical location of carbon emissions or absorption is of no consequence. What really matters is how much CO_2 can be sequestered (or emissions avoided), and different regions of the world are likely to provide quite different opportunities and comparative advantages in this regard. In many tropical countries, where combined biotic emissions appear to dwarf fossil fuel emissions, biotic sector policy options aimed at slowing deforestation, making agricultural production sustainable, and spurring reforestation seem promising in the short to medium term for mitigating global warming—so promising, in fact, that the United States might consider supporting the implementation of biotic policy options abroad as well as in the United States. This is the principle underlying implementation of the only forestry project funded specifically

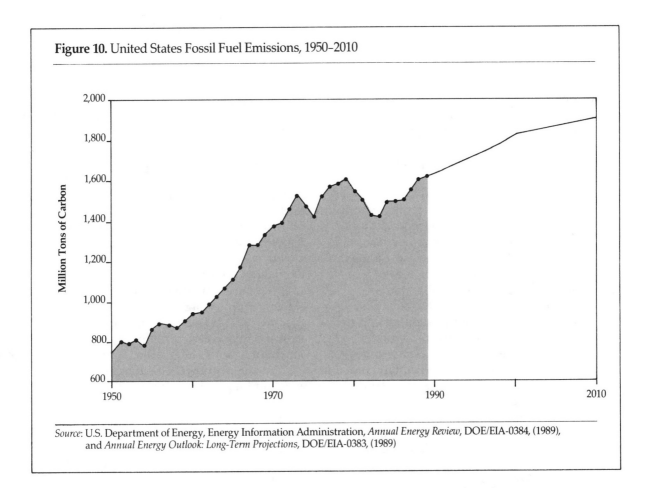

Figure 10. United States Fossil Fuel Emissions, 1950–2010

Million Tons of Carbon

2,000
1,800
1,600
1,400
1,200
1,000
800
600

1950   1970   1990   2010

Source: U.S. Department of Energy, Energy Information Administration, *Annual Energy Review*, DOE/EIA-0384, (1989), and *Annual Energy Outlook: Long-Term Projections*, DOE/EIA-0383, (1989)

as a carbon-emissions offset, a project that would have been prohibitively expensive in the United States. *(See Box 8.)*

2. Most biotic options are best used as part of a transitional energy strategy intended to combat global warming.

Biotic policy options are often seen as the flip side of emissions reductions. As a result, it is commonly assumed that a ton of carbon stored in trees is roughly comparable to a ton of carbon not emitted in the first place. But the complexities and uncertainties of biological systems argue against carrying this presumed equivalence too far.

First, trees withdraw large quantities of CO_2 from the air only as long as they are growing.

Consequently, physical and biological limits constrain the long-term value of large-scale conventional tree planting unless they can later be harvested and made into products without releasing a large pulse of carbon. Energy sector initiatives, in contrast, face no such constraint. Indeed, pursuing most biotic policy options today merely buys us time before energy sector initiatives have to be implemented anyway to eliminate permanently the emission of that quantity of carbon successfully being absorbed into biomass in the interim. Nevertheless, the value of buying 50 to 75 years to lessen the pain of switching to "smokeless" fuels that emit no CO_2 or to sustainable bioenergy systems that recycle CO_2 should not be underestimated.

Second, under most circumstances, years and even decades must pass before newly planted

47

Box 8. Offsetting U.S. Carbon Releases Abroad

Although planting trees to offset carbon emissions from fossil fuel consumption was first proposed in 1977, the idea was not acted upon until 1988, when Applied Energy Services (AES) began building a new 183-megawatt coal-fired powerplant in Connecticut.

AES, an independent power producer with a contract to sell power to Connecticut Power and Light, was concerned about how releasing approximately 15 million tons of carbon into the atmosphere, as the powerplant will over its 40-year life, would contribute to global warming. Approached for advice on how to neutralize the global warming impacts of the plant's CO_2 emissions, the World Resources Institute ultimately recommended that AES fund a sustainable agriculture and agroforestry project proposed for Guatemala by the relief organization CARE. Under the final terms of the agreement, AES contributed $2 million to the project, making it possible for CARE to attract additional financial and in kind support for the project valued at nearly $15 million.

WRI analysis concluded that the project will more than offset the 15 million tons of carbon emitted over the plant's lifetime. In addition, significant benefits will accrue to local residents. Over 10 years, the project will bring private and community woodlots; agroforestry for food, fuel, and fodder; and alley-cropping, live-fencing, and soil-conservation practices to about 40,000 Guatemalan farm families. If the processes put into action by the project last 40 years, the life span of the power plant, well over 100 million trees will be planted. Nevertheless, the most important carbon-storage benefit lies in protecting the region's existing standing forest. This protection is to be achieved by increasing the productivity of agricultural lands, thus reducing the need to destroy neighboring forest to secure fuelwood and make way for agriculture.

Source: Trexler, M.C., Faeth, P.E., and Kramer, J.M., 1989. *Forestry as a Response to Global Warming: An Analysis of the Guatemala Agroforestry and Carbon Sequestration Project.* World Resources Institute.

trees grow large enough to achieve the predicted carbon-storage goals. It is impossible to predict the extent to which natural change disasters, or human actions might reduce growth rates during this period or, for whatever reasons, tree mortality and decomposition might exceed growth. Although temperate forests are robust under current climate regimes, increased aridity owing to climate change could lead to lower regeneration and survival rates. Similarly, if tree mortality is heightened by increasingly severe fires, pest infestations, droughts, or weather, a significant fraction of the projected carbon storage benefits could simply fail to materialize. Overall, if significant climate change does occur within the United States, all of the forestry options described above may be needed just to keep biotic CO_2 releases from rising dramatically.

If significant climate change does occur within the United States, all of the forestry options described above may be needed just to keep biotic CO_2 releases from rising dramatically.

Given these uncertainties and constraints, it is not prudent to assume that all available policy options for mitigating global warming over the long term would be equally effective. Instead, biotic options might best be pursued as a transitional strategy to complement attempts to reduce fossil fuel emissions and not as a way to justify environmentally the combustion of fossil fuels in the long term. Yet, given the institutional, economic, and implementation barriers to rapidly reducing fossil fuel consumption, it makes sense to pursue the benefits of biotic policy options to the maximum extent feasible during the next several decades.

3. How biotic strategies are implemented will determine how significant ancillary benefits beyond global warming mitigation become.

Reducing the risk of global warming is not the only reason to plant trees. Tree planting in its various forms can protect fragile agricultural lands, conserve watersheds and aquifers, reduce erosion, preserve wildlife habitat, enhance recreation, make our cities more beautiful and comfortable, and further the nation's energy security. Because the carbon-sequestration benefits of many biotic options are by nature tentative and because the magnitude and rate of future global warming remain uncertain, there is every reason to maximize the ancillary benefits of whatever biotic policy initiatives are undertaken.

The type and quantity of ancillary benefits realized will depend on which biotic options are pursued and how. Although it might be simplest to plant millions of acres of new softwood plantations, there are other tree planting strategies that would promote worthy goals besides global warming mitigation. Planting trees along a river's edge, for example, can help control nonpoint source pollution and erosion while absorbing carbon. Planting mixed stands or even simulating natural forests can do far more to promote wildlife habitat than monoculture plantations can.

On the other hand, the ancillary benefits of tree planting should not be overstated. It has

recently been proposed, for example, that planting trees can reduce ozone, particulate, and sulfur dioxide pollution, thereby reducing the need for conventional pollution controls.[118] But though trees can filter some pollutants out of the air, absorbing significant amounts of pollution is likely to kill the trees.

4. Complex regulations and incentives must be accounted for when attempting to further carbon sequestration goals by directing or predicting private behavior.

Some biotic policy options involve the fine-tuning of current land-use and land-management practices by either private landowners or government land managers. Some farmers would become tree farmers for the first time; others would have to manage their stands more intensively. In any case, the difficulty of changing individuals' and agencies' priorities and missions should not be underestimated. Any attempt to alter the U.S. biotic carbon cycle must take hard reality into account: investment decisions about land use are quite sensitive to market events as well as to existing policies and subsidies that affect a landowner's bottom line, and these incentives are complex and sometimes contradictory. *(See Box 9.)*

Some of these incentives are also changing. The ownership and management of industrial timberlands, which now account for some 25 percent of all private timberlands, is a case in point. Many timberlands are being broken up and sold to developers and other interests, sometimes to fend off hostile takeovers, sometimes to pay off the debt engendered by such takeovers.[119] *(See Box 4.)*

5. Reliably increasing the size of U.S. biotic carbon sinks requires major changes in public policy.

The prognosis for increasing U.S. biotic carbon stores appears mixed. Programs such as the Conservation Reserve Program afford hope, but the decline of urban forests, the moribund state of rural forests in certain regions, the

Box 9. Private Decision-Making and Tree Planting

Various federal policies exist to foster reforestation, and federal, state, and private programs have been created to provide extension services, pest and fire control, and other types of support to tree growers. Although such government programs and policy decisions are not the only variables affecting the disposition of private lands, they can be quite important.

Federal land subsidies and price supports—major determinants of private land-use decisions—tend to shift as conditions change. Although the Soil Bank program of 1956 to 1962 was created to take fragile lands out of agricultural production and to reduce crop surpluses, for example, agricultural subsidies during the 1970s once again encouraged the conversion of fragile forest and wetlands to agricultural use. Today, many of these same acres are being withdrawn from agricultural use through such mechanisms as the Conservation Reserve Program. And by insuring up to 75 percent of normal crop yields even on lands commonly flooded, the USDA Federal Crop Insurance Corporation (FCIC) has implicitly encouraged the conversion of unsuitable lands to agricultural uses. The National Flood Insurance program also encourages the conversion of forested floodplains to agricultural and other uses.

Federal tax and insurance programs further complicate decision-making. Federal tax incentives—basically, expensing and amortizing the reforestation bills—cost the U.S. Treasury almost $500 million per year. At the same time, deductions for expenses incurred in clearing land have encouraged the conversion of forest land to other uses. Tax breaks on timber sales, property tax abatements that require harvesting, and

capital gains benefits for harvesting timber lands all have similar effects. One timely example includes the extension in 1986 of favorable tax treatment to the losses amassed by native corporations in Alaska. Unintentionally, those tax breaks set the stage for potentially large-scale deforestation in Alaska's southeastern forests.

The 1985 Farm Act, on the other hand, took a big step toward impeding the conversion of unsuitable lands to agricultural use and encouraging the removal of fragile lands from agricultural production. Under the bill's "sodbuster" provisions, farmers with highly erodible lands must take conservation measures on them or face the loss of USDA farm-program benefits. Parallel "swampbuster" provisions deny eligibility for program benefits to farmers who produce crops on wetlands converted to agricultural use after 1985—in effect, eliminating a major source of market demand for converted wetlands. By one estimate, the various provisions of the 1985 Act will ultimately prevent the conversion of almost 12 million acres of forest to agricultural land.

How these policies are implemented or revised will help determine how much carbon is stored, absorbed, or released from private U.S. lands.

Sources: Moulton, R.J. and Dicks, M.R., 1987. "Implications of the 1985 Farm Act for Forestry," in *Proceedings of the 1987 Joint Meeting of Southern Forest Economists and Midwest Forest Economists.* 1987, Asheville, North Carolina; Wells, K., 1990. "Endangered Again," *Wall Street Journal,* March 30; U.S. General Accounting Office, 1990. *Forest Service: Timber Harvesting, Planting, Assistance Programs and Tax Provisions.*

continuing sacrifice of woodlands and agricultural lands to urban sprawl and corporate needs for fast cash, and the harvest of many old-growth forests at an economic loss to the public treasury warrant increased concern over net carbon flows. Indeed, these and possibly other losses spurred by public and private natural-resource management decisions in the United States may be contributing more CO_2 to the atmosphere than the CRP and similar programs can take out of it.

Altering the U.S. biotic carbon cycle to help mitigate global warming will require much more than simply creating yet another federal program. Reducing net biotic emissions to zero and offsetting fossil fuel-related emissions of CO_2 are likely to require coordinating policy changes in many sectors. Even with the inclusion of the America the Beautiful initiative in the administration's 1991 budget, for example, the budget's net effect on future carbon stocks in U.S. forests may be negligible. *(See Box 6.)*

6. Problems are to be expected in determining what money spent on biotic policy buys.

The net carbon benefit to be gained by pursuing most biotic options can at best be determined only roughly. Calculating just the first-order carbon balance effects is often tricky, while second- and third-order effects can be hard to identify, much less quantify. Without such calculations to support them, however, biotic policy proposals run the risk of aggravating global warming as, for example, the proposed harvesting of old-growth forest to promote new tree growth on those lands well might. Other proposals might yield short-term benefits that are undone at least partially by their own second- and third-order consequences, as the subsidized afforestation of large areas of agricultural land in the United States is likely to do. Still others might be counteracted by forestry policies being pursued in other countries to combat global warming or achieve other aims. Or the carbon benefits of proposals could be undercut by government actions taken to solve problems other than

global warming. The failure of the United States over the last two growing seasons to produce as much wheat as it consumed suggests that lands now idled could well be returned to agricultural production. But some proposals could work as planned and store additional carbon. At any rate, properly assessing the likely carbon storage implications of any new policy will require a detailed but comprehensive analysis.

Fully and reliably accounting for all of these variables, of course, is impossible, since some of them reflect an inevitable uncertainty about the future. Careful ''carbon accounting'' should, however, provide as accurate as possible an assessment of first-order carbon balances. *(See Box 10.)* In addition, as many second-and third-order effects as possible should be incorporated into the analysis, particularly for projects large enough to affect timber and agricultural commodity prices. Even without firm projections, such an analysis should reveal how reliable the projections of primary benefits and cost-effectiveness are.

7. Implemented on a large scale, biotic policy options will not necessarily be more politically acceptable than alternative global-warming-mitigation measures.

Although planting trees to mitigate global warming is widely perceived as politically painless, the widespread implementation of biotic policy options could, in fact, prove quite contentious. Withdrawal of millions of acres of old-growth timberlands from harvesting would be at least a short-term blow to the timber sector in the Pacific Northwest. Greatly intensifying timberland management on remaining public and private lands would likely heat up debate over forest management practices and the loss of wildlife habitat. Taking a significant fraction of agricultural lands out of crop production would engender significant political opposition from well-entrenched interest groups. A large-scale switch to biofuels would significantly alter the energy production sector. In each case, important political and economic

Box 10. Performing a Comprehensive Carbon Accounting

For a rough idea of how much carbon a particular forestry or other biotic policy will sequester, several steps must be taken.

- Project realistic biomass growth rates, accounting for the species to be planted, likely soil conditions, and water availability. Also account for the possible impacts of climate change itself on water availability and growth rates at project sites.

- Project the proportion of biomass accumulation likely to be lost to natural forces, including fires, droughts, and insect or disease outbreaks.

- Account for carbon released during project implementation, including that in vegetation cleared from project sites, soil carbon released when lands are cleared or planted, and fossil fuel energy consumed during the planting, maintenance, and possible eventual harvesting on project sites. The energy consumed in producing fertilizers and pesticides for the project as well as that consumed in converting the biomass to liquid fuels or other energy forms, must also be accounted for.

- Account for any carbon that would have accumulated on the project site in the absence of the project.

- Account for the ultimate disposition of accumulated carbon. Biomass growth estimates are often premised on whole-tree biomass rather than just merchantable timber. If harvesting or thinning takes place, a significant proportion of the total carbon can be released back to the atmosphere—as roots, parts of trees left behind, or lumber wastes decompose.

- Avoid any double counting of carbon benefits. Delaying harvest of a forest containing one million tons of carbon, for example, can "reduce" emissions over time by only one million tons, regardless of how many years the delay is in force.

interest groups will be affected by the change. Considering the great uncertainties in projecting the global warming mitigation benefits of such biotic policy options, overcoming opposition from these economic and policy interests will be difficult.

As these general conclusions regarding the implementation of biotic policy options in the United States suggest, it is important for policymakers to differentiate between the theoretically possible and practically feasible global warming mitigation benefit associated with each biotic policy option. In the case of almost every option, the practical potential is considerably smaller than the theoretical potential. As Table 2 suggests, biotic options within the United States could in principle offset or displace

In the case of almost every option, the practical potential is considerably smaller than the theoretical potential.

almost all current fossil fuel CO_2 emissions, the equivalent of more than 1.5 billion tons of carbon per year. In practice, however, storing or displacing just 150 to 400 million tons of carbon per year, or 9 to 25 percent of current U.S. fossil fuel emissions, would require an aggressive and successful multipronged policy effort.

V. Recommendations

As Table 2 illustrates, estimates of the cost-effectiveness of the various biotic policy options do not unambiguously determine which options should be pursued first. Yet, given the comparative analysis of the various biotic options, some recommendations—the projected costs and carbon benefits of which are summarized in Table 3—do emerge from the foregoing analysis. If implemented, these recommendations could improve net U.S. carbon balances by 75 to 115 million tons per year, the equivalent of five to seven percent of current emissions. The estimated cost of implementing these recommendations, $525 million to more than $1.5 billion per year, compares favorably to a $1.1 to $2.5 billion estimate for achieving the same level of reductions in CO_2 emissions within the energy sector.[120] *(See Figure 11.)*

1. Selectively Convert Southeastern Pasture Land to Tree Cover.

An estimated 20 to 25 million acres of pasture land in the southeastern United States would be more economically productive if used to grow timber, probably pine. Converting this acreage from its current use appears to present the most cost-effective opportunity for significantly increasing the area of timberland in the United States and would also help control erosion and increase wildlife habitat. Conversion of these acres should be pursued before promulgating any new large-scale tree-planting initiatives that might require significant cost sharing and rental payments. Identifying the acres in question, providing the extension and technical assistance services required to educate landowners regarding alternative uses of their land, and possibly providing modest cost sharing or other mechanism to help plant the trees should result in the conversion of 15 million or more of the available acres at a relatively low cost. If it is assumed that an average allotment of $50 per acre would cover the various costs involved, a cash outlay of $750 million would be required to store 15 to 20 million tons of carbon annually. True societal costs, incorporating environmental benefits as well as future returns on planted timber, would be lower. Conversion of these areas should be a key aim of the president's America the Beautiful program, particularly given the modest sums currently budgeted for the program.

2. Expand Biomass-to-Electricity Demonstration Programs.

Any type of biomass can in principle be used to produce energy, and significant quantities of agricultural and woody wastes are already consumed for this purpose. But short-rotation woody and herbaceous crops represent the most cost-effective options for utilities and other large consumers of energy. Although biomass remains more expensive than coal on a MMBtu-delivered basis, if long-run world energy prices shift abruptly, so could these comparative costs. With oil at $30 per barrel, for example, intensive biomass cultivation for

Table 3. Recommended Carbon Storage Options

Recommendation	Projected Carbon Benefit[a] (millions of tons/yr)	Projected Costs (annualized) (millions of $)
1. Selectively Convert Southeastern Pasture Lands	15–20	$75–95
2. Expand Biomass-to-Electricity Demonstration Programs	20–30	$60–150
3. Improve Future Productivity of Harvested Timberlands	15–20	$150–400
4. Increase Tree Cover on CRP-Enrolled Acres	1–2	$10–25
5. Implement a Forestry Reserve Program	20–40	$200–1,000
6. Promote Urban Tree Planting	3–5	$30–75
7. Encourage Practices that Slow Carbon Loss and Build Soil Carbon	10–25	(not known)
8. Assess Existing Disincentives to Carbon-Storing Land Uses	NA	NA
9. Incorporate Carbon Storage Goals into the USFS Mandate	(not known)	(not known)
10. Improve Our Understanding of U.S. Carbon Stocks and Flows	NA	NA

NA = not applicable
a. The carbon benefit figure estimates carbon storage in biomass and in soils as well as fossil fuel carbon displaced through implementation of the biotic options.

energy production becomes economically viable on good sites. A demonstration program of 10 million acres would include these as well as more average sites, and it could be expected to displace almost two tons of fossil fuel carbon per acre currently, and almost three tons per acre as technologies advance. The cost per ton of net fossil fuel carbon displacement might reasonably range from $30 to $50 per ton, although incorporation of environmental and energy security values as well as any savings from commodity support programs would considerably lower any estimate of true societal costs.

The relatively little time needed to expand the use of biomass in electricity production and the relative reliability of the carbon displacement benefit of doing so justify a greater effort to determine this option's long-term technological, environmental, and economic feasibility. Demonstration efforts need to be directed not only at producing biomass (both woody and herbaceous) more efficiently but also at converting it more efficiently to electricity.

3. Improve Future Productivity of Harvested Timberlands.

A variety of forest management techniques can affect growth rates and other stand characteristics over the length of a growing cycle. The carbon-storage benefits of intensive management do not appear to justify the most costly

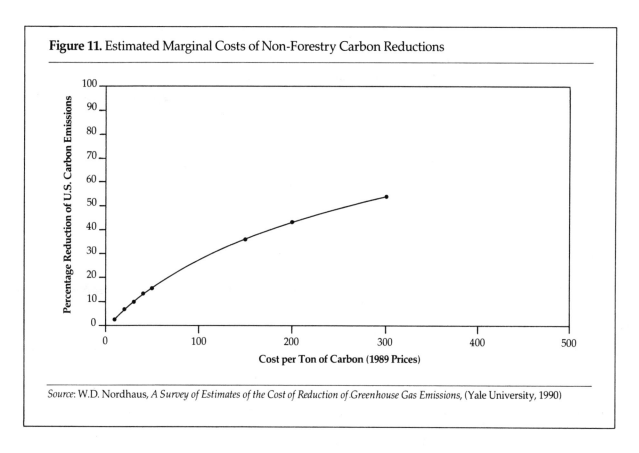

Figure 11. Estimated Marginal Costs of Non-Forestry Carbon Reductions

Source: W.D. Nordhaus, *A Survey of Estimates of the Cost of Reduction of Greenhouse Gas Emissions*, (Yale University, 1990)

and energy-intensive options—including forest fertilization and the use of genetically improved seedlings—and the restocking of growing stands. On some lands, however, modest silvicultural advancements could significantly increase growth rates and total carbon storage.

A great deal of intensification of timberland management, the equivalent of perhaps as much as 40 million tons of carbon per year, is already economically rational under existing market incentives. These acres, however, do not necessarily overlap with acres that will be harvested over the next couple of decades, and it is on these acres that intensive management is most appropriately begun. In addition, some of this intensification may result in simply favoring commercial over noncommercial biomass growth rather than actually increasing total carbon accumulation. Assuming that appropriate measures are taken on 30 to 40

million acres over the next couple of decades and an average incremental carbon storage of 0.5 tons per acre is achieved, 15 to 20 million tons of additional carbon might be stored annually. The average cost might well be $10 to $20 per ton, even if efforts are targeted at those forest acres where management intensification is judged already cost-effective. Long-term societal costs, once again, would likely prove lower.

4. Increase Tree Cover on CRP-Enrolled Acres.

Although some 34 million acres are currently enrolled in the Conservation Reserve Program, trees have been planted on only a little over 2 million acres. Because of restrictions on returning highly erodible lands to agricultural production, millions of additional and already enrolled acres could probably be converted to tree cover with relatively modest changes to the relevant CRP contract terms and lengths.

The USDA's Economic Research Service has estimated that over one million acres could be converted to tree cover at a cost of $100 to $200 million, although better outreach and extension services to affected farmers might well lower this cost estimate. The long-term nature of the tree cover would help prolong the environmental benefits of having removed this land from agricultural production, thus significantly cutting the societal cost of the option.

5. Implement a Forestry Reserve Program.

Based on the model provided by the Conservation Reserve Program, an FRP would offer the opportunity to identify less economically valuable land and encourage its conversion to tree cover. An FRP would rely on the same type of cost sharing and rental payment provisions found in the CRP, though it would involve lower per-acre costs. A target of creating an FRP of 20 million acres over the next decade (not counting the 15 million acres discussed in recommendation No. 1) would appear reasonable because many tens of millions of acres of economically or environmentally marginal crop and pasture lands are available. At a cost of $10 to $30 per ton of carbon stored, such a program would cost between $200 million and $1 billion per year. Because of uncertainties regarding future timber prices and the costs of future commodity support programs, this cost estimate does not account for financial returns or savings in either area.

In a new Forestry Reserve Program, however, the U.S. forestry community is likely to want to focus on what it already does best, which is pursue conventional industrial plantation forestry. This natural inclination should be resisted. If carried out on environmentally robust lands, land conversions under a Forestry Reserve Program might have few benefits besides carbon storage. If carried out on environmentally sensitive lands, however, the same magnitude of effort might significantly reduce erosion and nonpoint-source water pollution. If carried out in innovative ways even on nonsensitive lands, perhaps through the

construction of windbreaks and riparian corridors, such conversions would still promote a variety of environmental and policy goals in addition to carbon storage.

Before public funds are spent withdrawing lands from pasture or other alternative uses, the current policies that encourage these land uses should be reviewed and modified with global warming concerns in mind.

In any case, before public funds are spent withdrawing lands from pasture or other alternative uses, the current policies that encourage these land uses should be reviewed and modified with global warming concerns in mind.

6. Promote Urban Tree Planting Widely and Vigorously.

As noted, urban trees can both store carbon and reduce the cooling and heating requirements of homes and businesses, thus displacing sizable quantities of fossil fuels. Urban forestry initiatives should be directed primarily at urban areas where energy use for cooling is already significant or is likely to become significant soon. Through a combination of governmental and nongovernmental tree-planting and tree-protection initiatives (including changes in local zoning and development requirements), it should be possible to reverse the ongoing decline of urban forests and prepare cities for the potentially harsh effects of future global warming. Rising urban temperatures, for example, will aggravate the urban heat-island effect, favor the formation of such damaging pollutants as tropospheric ozone, and generally increase stress on trees. Although the total potential of urban tree-planting programs to mitigate global warming is limited, it combines the advantages of cost-effectiveness and multiple ancillary benefits. It may also help reduce the need to increase utility capacity to handle peak summer loads, and it should benefit from considerable public goodwill and volunteer efforts.

If targeted at the appropriate areas of the United States, an expanded urban forestry program might store or displace three to five million tons of carbon annually at an average societal cost of $10 to $15 per ton. Implementation of a nontargeted program or incentive structure, however, would result in considerably fewer carbon displacement benefits and higher costs.

7. Encourage Agricultural Practices that Slow Rates of Soil Carbon Loss and Enhance Soil Carbon Buildup.

Many agricultural and other land-use practices favor the release of soil carbon rather than its retention or augmentation. Clearing and farming forested lands generally reduces soil carbon by up to 40 percent, and conventional monoculture tillage and harvesting practices work against the restoration of the lost soil carbon. More than 200 million tons of soil carbon were lost from farm and pasture acreage annually in the United States in the early 1980s, an unknown portion of which actually oxidized to the atmosphere as CO_2.

With modern sustainable agricultural practices that are often already economically profitable in their own right, it should be possible to build up organic matter in the soil by a small amount per year for a decade or more. On 300 million acres of agricultural lands, an increase of just 0.25 tons of carbon per acre per year could increase carbon storage by 75 million tons annually, in addition to whatever quantity of carbon oxidation is avoided by slowing erosion. If the true societal costs of erosion and the other environmental impacts often associated with conventional agricultural practices are as high as is often estimated, major programs to stabilize and increase soil carbon levels are already economically justified. On a cost-per-ton-of-carbon basis, the societal

costs of this option likely range from negative to modest. Considering the political and bureaucratic hurdles to its widespread adoption, however, it seems unrealistic to expect to increase biotic carbon storage by more than 10 to 25 million tons of carbon per year through this mechanism in the near term.

8. Assess Existing Disincentives to Carbon-Storing Land Uses.

Diverse governmental programs and rules and regulations influence land-use and land-management decisions in the United States— among them, federal commodity support programs, federal timber sale programs, federal tax and insurance policies, state real estate taxes, and local zoning regulations. Such programs and incentives should be reviewed with cost-effective ways to sequester or displace additional carbon in mind, and programs that counteract these goals should be rooted out. It makes little sense to reforest millions of acres of agricultural land to combat global warming at significant expense if nothing is done to slow the rate of loss of carbon from forests and other lands.

9. Incorporate Carbon Storage Goals in the Mandates of U.S. Land Management Agencies.

The United States Forest Service and other land management agencies must take considerable responsibility in coming years for helping protect U.S. carbon reservoirs and for preparing forest management strategies for a warmer future. As a first step, carbon management and storage should be added to the Forest Service's multiple-use mandate. More specifically, timber-rotation lengths should be selected on the basis of both carbon storage and economic returns, and the practice of determining economically appropriate harvest levels on the basis of production targets, rather than vice versa, should be ended.[121] Greater emphasis on carbon management within the USFS and other agencies might also result in additional protection of old-growth forest, reduction in the reforestation backlog, the elimination of harvesting (particularly if below-cost) in areas likely to regenerate poorly or slowly, and extensive research on adapting forest management strategies to the threat posed by global warming. Given the subtle nature of global warming, the Forest Service will probably also have to take primary responsibility for developing the knowledge base that private timberland managers need to adjust their own adaptive strategies.

As part of government agencies' review of their possible responses to global warming in the forestry sector, analysis should be undertaken to assess the potential for increasing forest cover on public lands. Although agencies such as the Forest Service and the Bureau of Land Management contend that few planting opportunities are available on public lands, so much land is under the control of these and other public agencies that this conclusion appears excessively pessimistic.

10. Improve Our Understanding of U.S. Carbon Stocks and Flows.

Considerable research is needed to improve the current understanding of U.S. carbon stocks and trends both on and off forested lands and to improve our ability to characterize the potential biomass productivity of non-forested lands. Available forestry data tend to be biased toward commercial biomass, and important elements of the carbon cycle are therefore overlooked in existing estimates. Such a data gathering and modeling effort is likely to be beyond the scope of any single natural resource management agency (such as the USFS) and may well require a broader interagency approach. Making the model and its underlying data base publicly available would make public consideration and review of biotic policy options much more practical.

Mark C. Trexler is an Associate with the WRI Program in Climate, Energy and Pollution, and directs the program's work on domestic and international forestry as a response to global warming. Before joining WRI, Dr. Trexler worked with such organizations as the International Union for the Conservation of Nature and the California Energy Commission on environmental and energy policy issues.

Notes

1. Houghton, R.A., 1990. ''Emissions of Greenhouse Gases,'' in Myers, N., ed., *Deforestation Rates in Tropical Forests and Their Climatic Implications.* Friends of the Earth.

2. It has been estimated that deforestation in 1987 contributed more than 3 billion tons of carbon to the atmosphere, more than 1.3 billions tons of it from Brazil alone. World Resources Institute in collaboration with United Nations Environment Programme and United Nations Development Programme, 1990. *World Resources 1990–91.* Oxford University Press. A more recent estimate, however, argues that these estimates overstate Brazilian deforestation by as much as a factor of four. Fearnside, P.M., et al., 1990. *Deforestation Rate in Brazilian Amazonia.* Brazilian National Secretariat of Science and Technology.

3. Houghton, R.A., et al., 1983. ''Changes in the Carbon Content of Terrestrial Biota and Soils Between 1860 and 1980: A Net Release of CO_2 to the Atmosphere,'' *Ecological Monographs* 53:235–62.

4. Opinion is deeply split among scientists as to whether such fertilization is already occurring. Foresters note their inability to observe rising growth rates in the world's forests. Some carbon modelers, however, believe that a CO_2 fertilization effect must be assumed in order to balance the global carbon cycle and that such an effect would not necessarily be measurable at the individual stand level. For a discussion of the science underlying the carbon fertilization debate, *see* Kramer, P.J., and Sionit, N., 1987. ''Effects of Increasing Carbon Dioxide Concentration on the Physiology and Growth of Forest Trees,'' in Shands, W.E., and Hoffman, J.S., eds. *The Greenhouse Effect, Climate Change, and U.S. Forests.* Conservation Foundation.

5. Houghton, R.A., 1990. ''The Future Role of Tropical Forests in Affecting the Carbon Dioxide Concentration of the Atmosphere,'' *Ambio* 19(4):204.

6. Intergovernmental Panel on Climate Change, 1990. *Policymakers Summary: The Formulation of Response Strategies.* World Meteorological Organization and United Nations Environment Programme.

7. U.S. Department of Agriculture, Forest Service, 1990a. *An Analysis of the Timber Situation in the United States: 1989–2040.* Editorial draft.

8. Houghton estimates the carbon content of temperate grasslands after disturbance for cultivation at about 42 tons per acre as soil carbon, with an additional couple of tons of carbon in vegetation. (Houghton, R.A., et al., 1983.) Using these figures, one could estimate the minimum carbon content of crop and pasture lands at 45 billion tons of

carbon. Birdsey, however, believes that U.S. forest soils average more than 46 tons of carbon per acre, for a total of some 35 billion tons. (Birdsey, R.A., in press. "Potential Changes in Carbon Storage Through Conversion of Lands to Plantation Forests," in *Proceedings, North American Conference on Forestry Responses to Climate Change,* 1990, Washington, D.C. Climate Institute.) Because the average carbon content of forest soils should be considerably higher than that of crop and pasture land soils, it seems likely that Houghton's number is significantly too high or Birdsey's is significantly too low. Birdsey, for example, estimates that grasslands probably hold some 25 tons per acre. (Birdsey, R.A., pers. comm.) In either case, forest ecosystems are likely by far to exceed agricultural lands as carbon repositories.

9. Estimates of the vegetational and soil carbon content of an acre of old-growth biomass in the Pacific Northwest range up to almost 300 tons per acre. Estimates of tropical forest biomass in Latin America are less than 100 tons per acre even for undisturbed productive forest. Harmon, M.E., Ferrell, W.K., and Franklin, J.F., 1990. "Effects in Carbon Storage of Conversion of Old-Growth Forests to Young Forests," *Science* 247:699–702; Brown, S., and Lugo, A.E., 1984. "Biomass of Tropical Forests: A New Estimate Based on Forest Volumes," *Science* 223:1290–93.

10. *See* note 4. Also *see* Tans, P.P., Fung, I.Y., and Takahashi, T., 1990. "Observational Constraints on the Global Atmospheric CO_2 Budget," *Science* 247:1431–38.

11. Birdsey, R.A., in press.

12. Hansen, E.A., in press. "Biological Opportunities to Increase Tree Biomass Accumulation and Yield from Timberland," in *Proceedings, North American Conference on Forestry Responses to Climate Change,* 1990, Washington, D.C. Climate Institute.

13. Because trees are unlikely to decompose and oxidize to CO_2 in the same year they die, it can be argued that they do not constitute a source of carbon for the year and that total mortality should not be subtracted from net growth in arriving at an overall carbon balance. On the other hand, some fraction of the biomass in trees that died in every previous year will oxidize during the year in question. Rather than attempt to calculate the fraction of mortality from each previous year oxidizing in the present year, it is much simpler for carbon accounting purposes simply to count total nonharvested mortality in a given year against total growth in that same year.

14. Birdsey, R.A., pers. comm.

15. Shands, W.E., and Hoffman, J.S., eds., 1987.

16. Davis, M.B., 1989. "Lags in Vegetation Response to Greenhouse Warming," *Climatic Change* 15(1–2):75–82; Smith, J.B., and Tirpak, D., eds., 1989. *The Potential Effects of Global Climate Change on the United States: Report to Congress.* U.S. Environmental Protection Agency, Office of Policy, Planning and Evaluation.

17. Leverenz, J.W., and Lev, D.J., 1987. "Effects of Carbon Dioxide-Induced Climate Changes on the Natural Ranges of Six Major Commercial Tree Species in the Western United States," in Shands, W.E., and Hoffman, J.S., eds.

18. Sandenburgh, R., Taylor, C., and Hoffman, J.S., 1987. "Rising Carbon Dioxide, Climate Change, and Forest Management: An Overview," in Shands, W.E., and Hoffman, J.S., eds.

19. This fact is sometimes cited to argue that human activities are responsible for just a small part of the accumulation of additional CO_2 in the atmosphere. The argument is highly misleading, however, for in the

absence of human activities, natural systems normally would emit and absorb roughly equal amounts of CO_2, thus maintaining a planetary balance. It is the perturbation to the CO_2 cycle introduced by human activities that accounts for the rise in atmospheric CO_2 concentrations.

20. Trees pump water up from their roots to their leaves, where it evaporates into the air, moving up to 100 gallons of water per day. The evaporation of this water absorbs heat from the air and lowers surrounding air temperatures.

21. An albedo of one refers to a surface that completely reflects all light, while a value of zero refers to one that completely absorbs the energy in light. Trees have an albedo of around 0.25. A practical upper-limit estimate of the albedo of a highly vegetated southern city with light-colored surfaces on its pavement and buildings is 0.40. Although trees can increase albedo under some circumstances, on average the addition of trees to the urban environment probably reduces rather than increases albedo, thus offsetting some of their global warming benefits. Akbari, H., et al., 1989. "Saving Energy and Reducing Atmospheric Pollution by Controlling Summer Heat Islands," in *Controlling Summer Heat Islands*. Proceedings of the Workshop on Saving Energy and Reducing Atmospheric Pollution by Controlling Summer Heat Islands, Lawrence Berkeley Laboratory, University of California, Berkeley.

22. Akbari, H., et al., 1989.

23. Representative cities where the effects of urban forestry have been modeled include Sacramento, California; Lake Charles, Louisiana; and Phoenix, Arizona.

24. The three main effects measured in the computer model were reduced solar gain (shading), evapotranspiration (evaporation of water by vegetation), and reduced wind speed. Even when trees were strategically placed to maximize shading, evapotranspiration accounted for 65 to 90 percent of the energy savings; only 10 to 30 percent was attributable to shading itself. Akbari, H., et al., 1989; Huang, Y.J., et al., 1986. *The Potential of Vegetation in Reducing Summer Cooling Loads in Residential Buildings.* Lawrence Berkeley Laboratory, University of California.

25. Huang, Y.J., et al., 1986.

26. Moll, G., 1989. "The State of Our Urban Forest," *American Forests*, November/December:61–64.

27. Robinette, G., 1977. *Landscape Planning for Energy Conservation.* Environmental Design Press, prepared by the American Society of Landscape Architecture Foundation.

28. The AFA also estimates that an additional 60 million planting sites are available along streets and 160 million in parks and other open places in urban areas.

29. Akbari, H., et al., 1989.

30. Akbari, H., et al., 1989. This estimate does not separate out the costs of tree planting and increasing albedo.

31. Moll, G., 1989.

32. The other required precursors to smog, however, are oxides of nitrogen (NO_x), which are produced primarily through the combustion of fossil fuels. Chameides, W.L., et al., 1988. "The Role of Biogenic Hydrocarbons in Urban Photochemical Smog: Atlanta as a Case Study," *Science* 241:1473.

33. *Weekly Bulletin*, June 5, 1989. Environmental and Energy Study Institute, Washington, D.C.

34. Economic Incentives Committee, 1989. *Incentives for Tree Planting in the United*

States to Slow the Build-up of Atmospheric Carbon Dioxide. Report of the ''Trees for U.S.'' Task Force, EIC.

35. U.S. Department of Agriculture, Economic Research Service, 1989. *Agricultural Resources: Cropland, Water and Conservation: Situation and Outlook Report*. USDA/ERS AR–16.

36. U.S. Department of Agriculture, Forest Service, 1990b. *America the Beautiful: National Tree Planting Initiative*.

37. Parks, P.J., in press. ''Potential for Converting Marginal Crop and Pasture Lands to Forests,'' in *Proceedings, North American Conference on Forestry Responses to Climate Change*, 1990, Washington, D.C. Climate Institute; U.S. Department of Agriculture, Forest Service, 1990b.

38. These regions are defined as states east of the Mississippi River plus Minnesota, Missouri, Texas, Oklahoma, Arkansas, and Louisiana. Economic Incentives Committee, 1989.

39. Parks, P.J., in press; U.S. Department of Agriculture, Forest Service, 1988a. *The South's Fourth Forest: Alternatives for the Future*. Forest Resources Report No. 24.

40. Marland, G., 1988. *The Prospect of Solving the CO_2 Problem Through Global Reforestation*. DOE/NBB-0082, TRO39. U.S. Department of Energy, Office of Basic Energy.

41. These figures are based on total tree carbon productivity per acre rather than just the carbon content of merchantable timber. The multiplier for converting from merchantable timber to total tree biomass can vary from 1.5 to over 2.5 depending on the species. A conservative value of 1.6 was used in these estimates. This multiplier does not account for carbon that might accumulate over time in understory vegetation, in downed branches and other litter, and in soils.

Biomass multipliers intending to convert merchantable timber estimates into total ecosystem carbon in mature forests are estimated to be as high as 7.7, although the applicability of these high multipliers to conventional tree-planting projects remains unclear. Moulton, R.J., and Richards, K.R., 1990. *Costs of Sequestering Carbon Through Tree Planting and Forest Management in the United States*. U.S. Department of Agriculture, Forest Service.

42. Birdsey, R.A., pers. comm.

43. Wood removed during thinning must also be netted out if it is not incorporated into long-lasting wood products or compensated for by sufficiently better growth in the remaining trees.

44. The choice of a time line is of great importance to the estimation of the carbon benefits of planting trees that are to be harvested. If assessed in year 50 of a 50-year rotation, for example, carbon benefits will be at their peak. If assessed at year 51 just after harvest, carbon benefits could well be significantly less than what they were in year 50. If assessed at year 75, the carbon benefit (new growth + long-term product from the first rotation) may be approaching that seen prior to harvest. If assessed at year 99, the carbon benefit for the harvested acre could exceed that for a nonharvested acre.

45. Brandle, J.R., Wardle, T.D., and Bratton, G.F., forthcoming. *Shelterbelt Opportunities and the Potential Impacts on Global Warming*. American Forestry Association.

46. These savings would be realized by withdrawing the planted acres from intensive agricultural production, although total agricultural productivity is not projected to fall. They also accrue from reducing energy consumption in homes and farmsteads and from reducing snow-removal requirements. Brandle, J.R., Wardle, T.D., and Bratton, G.F., forthcoming.

47. Faeth, P., pers. comm.

48. Economic Incentives Committee, 1989.

49. National Research Council, 1989. *Alternative Agriculture.* National Academy Press.

50. Government expenditures on farm programs differ from year to year based on many variables and have ranged from $10 to $25 billion per year over the last decade. The total over the last five years was $93 billion. U.S. Department of Agriculture, 1989a. *Agricultural Handbook #684*, Economic Research Service.

51. Cubbage, F.W., forthcoming. *Current Federal Land Conversion Programs: Accomplishments, Effectiveness, and Efficiency.* American Forestry Association.

52. Economics Incentives Committee, 1989.

53. Ibid.

54. Ibid. These cost estimates exceed those suggested for CRP land conversions because of the much larger scale of the program in the face of rising marginal costs. Convincing farmers with their land already enrolled in the CRP program to plant trees may well be cheaper than convincing those still using their land for commercial crop production.

55. These carbon storage figures incorporate high estimates of biomass accumulation other than actual carbon accumulation in wood. (*See* note 41.) Moulton, R.J., and Richards, K., 1990.

56. It is true that falling timber prices should increase the demand for timber products and possibly carbon storage in wood products, but one would still expect a net decline in investment on existing timberlands.

57. U.S. Department of Agriculture, 1989b. *Conservation Reserve Program: Progress Report and Preliminary Evaluation of the First Two Years.*

58. U.S. Department of Agriculture, Agricultural Stabilization and Conservation Service, 1990. *Conservation Reserve Program "Logo Package."*

59. Analysis of the FIP and the ACP has found that conversions to forest cover under those programs has proven cost-effective for the landowners, with internal rates of return varying from 10 to 15 percent over the last 15 years. Cubbage, F.W., forthcoming.

60. Alig, R.J., et al., 1980. "Most Soil Bank Plantings in the South Have Been Retained; Some Need Follow-Up Treatments," *Southern Journal of Applied Forestry* 4(1):60–64.

61. Kurtz, W.B., et al., 1980. "Retention and Condition of Agricultural Conservation Program Conifer Plantings," *Journal of Forestry* 78(5):273–76.

62. Vasievich, J.M., 1983. "Constraints and Risks: Physical Losses/Risks." *Proceedings of the American Forestry Association 108th Annual Meeting*, 1983, Washington D.C. American Forestry Association.

63. Royer, J.P., and Moulton, R.J., 1987. "Reforestation Incentives," *Journal of Forestry* 85:8.

64. Sedjo, R.A., and Lyon, K.S., 1990. *The Long-Term Adequacy of World Timber Supply.* Resources for the Future.

65. Because a significant fraction of global wood production is from natural forests, a reduction in commercial wood harvesting cannot be directly equated to a decrease in standing biomass. Some proportion of the forgone eight billion cubic feet in commercial wood production outside the new tropical plantations, for example, is likely to be the result of leaving standing forest alone (not a net carbon loss) rather than forgoing

plantations that otherwise would have been grown (a net carbon loss).

66. U.S. Department of the Interior, Bureau of Land Management, 1987. *Public Land Statistics 1946–1986.*

67. U.S. Department of Defense, 1990. *Our Nation's Defense and the Environment: A Department of Defense Initiative.*

68. U.S. Environmental Protection Agency, 1989a. *Reforest America: A Cost Effective Program to Reduce U.S. CO_2 Emissions by 10 Percent.*

69. U.S. Environmental Protection Agency, 1989b. *Policy Options for Stabilizing Global Climate. Report to Congress.* Draft.

70. Although the National Forest Management Act of 1976 mandated that the Forest Service eliminate its reforestation backlog by 1985, subsequent congressional investigations revealed that most of the backlog was eliminated through administrative means rather than by actual reforestation. A 1985 report to the Committee on Appropriations by its Surveys and Investigations Staff revealed that the Forest Service sometimes claims replanting of backlog acreage even when no tree planting is planned, before planned planting has occurred, or when only a part of an area is in fact treated. The report concluded that large portions of the backlog acres were removed from the rolls not by reforestation but by administrative fiat reclassifying the land out of the timberland system. Surveys and Investigation Staff, 1985. *10-Year Reforestation Backlog Elimination Program of the U.S. Forest Service.* A Report to the Committee on Appropriations, U.S. House of Representatives; Carey, H.H., et al., 1989. *National Forests: Policies for the Future*, vol. 3, *Reforestation Programs and Timberland Suitability.* Wilderness Society.

71. Wubbels, D., pers. comm.

72. This modest estimate of land availability is generally consistent with the AFA's estimate that an additional 60 million trees could be added along public roadways.

73. Harmon, M.E., Ferrell, W., and Franklin, J.F., 1990.

74. Timber sales are characterized as below cost when the Forest Service spends more on sale preparation and administration, mitigation, road construction, and reforestation than it receives for the sold timber. In the case of the Tongass National Forest in Alaska, for example, it is estimated that $32 million was received in timber sales revenue but some $287 million was spent on the sales. Below cost sales are estimated to occur in at least 76 national forests and to involve some 125 million acres of national forest lands. Wilderness Society, 1989. *New Directions for the Forest Service: Instructions from the President of the United States to the Chief of the Forest Service (Recommendations of the Wilderness Society)*; Repetto, R., and Gillis, M., eds., 1988. *Public Policies and the Misuse of Forest Resources.* Cambridge University Press.

75. Net carbon flows will depend on the biology of the forest, including the amount of standing timber and the likelihood and speed of its regeneration. It will also depend on the disposition of harvested biomass.

76. Wilderness Society, 1989.

77. For a discussion of the controversy over forest management practices, see Robinson, G., 1988. *The Forest and the Trees: A Guide to Excellent Forestry.* Island Press.

78. It is estimated that at least 80 percent of the old-growth forest in national forests remains potentially available for logging. (Wilderness Society, 1989). This situation may change in response to the listing of the northern spotted owl as a threatened species under the Endangered Species Act.

79. This estimate is not necessarily the equivalent of a doubling of total "carbon growth" on U.S. timberlands because an increase in merchantable timber growth might well imply a decrease in nonmerchantable species and growth.

80. U.S. Department of Agriculture, Forest Service, 1990b; U.S. Congress, Office of Technology Assessment, 1983. *Wood Use, U.S. Competitiveness and Technology*. OTE-E-210.

81. Wilderness Society, 1989.

82. Current commercial timber accumulation on an average acre of timberland totals less than 1 ton of carbon per year, so an increase of 0.5 to 1 ton per acre would represent a significant advance over current growth rates.

83. The lower end of the range assumes the interim loss to the atmosphere of roughly 200 tons of carbon from the cutting of one acre of old-growth forest, and the upper end of the range assumes the long-term loss of some 135 tons of carbon. *(See Box 3.)*

84. Judis, J.B., 1990. "Ancient Forests, Lost Jobs Ride Wings of Spotted Owl," *In These Times*, August 1–14, 1990. Institute for Public Affairs, Chicago, Illinois.

85. Winjum, J., pers. comm.

86. Some analysis suggests that 75 percent of the cost of timberland improvement that is not already prima facie economically justified will be made up for in the economic value of the harvested product, considerably reducing the projected net cost of the option. Without understanding the second-order effects of the increased management and planting on long-term timber supplies and markets, however, such an assumption is risky.

87. Because old-growth trees survive on a tight energy budget, shifts in climate may stress and weaken these trees more than younger specimens. Scientists have concluded that "[a]n increase in respiration leading to a loss of vigor in old growth stands could have devastating effects in some forest regions, particularly in the West and in parts of the East." U.S. Department of Agriculture, Forest Service, 1988b. *Forest Health and Productivity in a Changing Atmospheric Environment: A Priority Research Program*.

88. Birdsey, R.A., in press.

89. Row, C., and Phelps, R.B., in press. "Carbon Cycle Impacts of Improving Forest Products Utilization and Recycling," in *Proceedings, North American Conference on Forestry Responses to Climate Change*, 1990, Washington, D.C. Climate Institute.

90. Harmon, M.E., et al., 1990. Sedjo, however, suggests using a rate as low as 25 percent. Sedjo, R.A. and Solomon, R.A., 1989. *Climate and Forests*. Resources for the Future.

91. For a discussion of biomass utilization technologies, see U.S. Department of Energy, 1988. *Five Year Research Plan 1988–1992; Biofuels: Renewable Fuels for the Future*, DOE/CH10093-25. Biofuels and Municipal Waste Technology Program; Brower, M., 1990. *Cool Energy: The Renewable Solution to Global Warming*. Union of Concerned Scientists; Rader, N., 1989. *Power Surge: The Status and Near-Term Potential of Renewable Energy Technologies*. Public Citizen; Flavin, C., and Piltz, R. 1989. *Sustainable Energy*. Renew America.

92. Because it is projected that we will run out of liquid fossil fuels at some point in the future, displacing the production of a barrel of oil on the margin today does not mean the permanent sequestration of that amount of carbon. What it does do is delay the release of that carbon by the amount of time between the original substitution and

the ultimate depletion of supplies of that fossil fuel. Should depletion of that fossil fuel source be more than a century away, there is no conceptual difference between the global warming implications of burning one ton of carbon in the form of biocrude gasoline made from water hyacinths and storing the same amount of carbon in a long-lived Douglas fir tree.

93. Productivity figures are expressed in terms of tons of carbon per acre per year to maintain consistency with earlier discussion. These figures are derived by multiplying dry harvested biomass weights by 0.5. The conversion to carbon is not intended to suggest that no other sources of energy (e.g., hydrogen) are present in the biomass. Wright, L.L., Cushman, J.H., and Layton, P.A., 1989. "Expanding the Market by Improving the Resources." *Biologue* 6(3):12–19. National Wood Energy Association.

94. Wright, L.L., Graham, R.L., and Turhollow, A.F., in press. "Short-Rotation Wood Crop Opportunities to Mitigate Carbon Dioxide Buildup," in *Proceedings, North American Conference on Forestry Responses to Climate Change*, 1990, Washington, D.C. Climate Institute; InterTechnology Corporation, 1975. *The Estimated Availability of Resources for Large-Scale Production of SNG by Anaerobic Digestion of Specially Grown Plant Material: Final Report.* American Gas Association Project No. IU 114–1.

95. Estimates of the energy inputs and outputs involved in ethanol production from feedstocks such as corn can vary depending on the accounting method used and credits given for secondary products such as cattle and chicken feeds. Pimentel, D., et al., 1988. "Food Versus Biomass Fuel: Socioeconomic and Environmental Impacts in the United States, Brazil, India, and Kenya," in Chichester, C.O., and Schweigert, B.S., eds., *Advances in Food Research*, vol. 32. Academic Press, Inc.

96. For every 1.4 Btus of SNG produced, 1.0 Btu of conventional fuel would be required to grow the crop and run the conversion facility. InterTechnology, 1975.

97. Wright, L.L., Graham, R.L., and Turhollow, A.F., in press.

98. InterTechnology, 1975.

99. U.S. Department of Energy, 1988.

100. Brower, M., 1990.

101. Wright, L.L., Graham, R.L., and Turhollow, A.F., in press.

102. U.S. Department of Energy, 1988.

103. Current DOE research in plant-oil-derived diesel fuel includes the development of microalgae systems in the desert Southwest, where there is high incident solar radiation, plenty of land, and large reservoirs of saline water. To lower the cost of fuel production, DOE is looking for strains of microalgae that exhibit high oil accumulation and rapid growth. In fact, researchers have identified strains of microalgae that produce up to 65 percent of their body weight in oil. Genetic engineering will be used to maximize oil production because the species that have high lipid yields generally have low growth rates. The viability of large-scale production of oils is also being addressed as researchers verify the availability of saline waters and affordable supplies of CO_2. U.S. Department of Energy, 1988.

104. U.S. Department of Energy, 1988.

105. Ostlie, L.O., 1989. "The Whole Tree Burner: A New Technology for Power Generation," *Biologue* 5(3):79.

106. These prices precede any long-term impacts of Iraq's invasion of Kuwait in August 1990.

107. Various studies show higher injury and illness rates in the agricultural and forestry sectors than in the fossil fuels sector per unit of energy produced. U.S. Congress, Office of Technology Assessment, 1980. *Energy from Biological Processes,* vols. 1 and 2; Pimentel, D., et al., 1988.

108. Wright, L.L., Graham, R.L., and Turhollow, A.F., in press.

109. Biomass gasification is one alternative receiving considerable attention. Williams, R.H., 1989. Statement before the House Appropriations Committee, Subcommittee on Foreign Operations. February 21.

110. Ostlie, L.O., 1989.

111. U.S. Department of Agriculture, Soil Conservation Service, 1987. *Natural Resources Inventory.* Statistical Bulletin No. 756; Pimentel, D., et. al., 1988.

112. This reflects an increase in soil organic matter of 0.02 percent per year. Rader, N., 1989.

113. U.S. General Accounting Office, 1990. *Alternative Agriculture: Federal Incentives and Farmers' Opinions.* Program Evaluation and Methodology Division.

114. Soil losses in the southern United States during the past 100 to 200 years have reportedly caused crop-biomass reductions ranging from 25 to 50 percent for a variety of crops. Pimentel, D., et al., 1988.

115. Faeth, P., and Repetto, R., in press. *The Economics of Sustainable Agriculture.* World Resources Institute.

116. Edmonds, J.A., et al., 1989. *A Preliminary Analysis of U.S. CO_2 Emissions Reduction Potential from Energy Conservation and the Substitution of Natural Gas for Coal in the Period to 2010.* U.S. Department of Energy.

117. Congressional Budget Office, 1990. *Carbon Charges as a Response to Global Warming: the Effects of Taxing Fossil Fuels.*

118. Illinois Commerce Commission, 1989. *The Illinois Air Quality Enhancement Program Through Comprehensive Tree Planting: Program Prospectus.*

119. Porterfield, A., 1989. ''Railroaded: The LBO Trend on Wall Street is Playing Havoc with that Nation's Forests,'' *Common Cause Magazine,* September/October: 21–23.

120. Derived from Nordhaus, W.D., 1990. A Survey of Estimates of the Cost of Reduction of Greenhouse Gas Emissions. Yale University.

121. Repetto, R., and Gillis, M., eds., 1988.

Appendix
Source or Sink? A Preliminary Analysis of Carbon Flows in the United States

The carbon cycle of the United States is poorly understood from a quantitative standpoint, partly because impacts on the carbon cycle have not figured heavily in natural resource management decisions. Indeed, swiftly changing and sometimes contradictory public policies, a basic lack of biological information on long-term resource trends, and the unknowns surrounding the potential impacts of such exogenous variables as global warming and acid rain suggest the need for great caution in estimating the effects and cost-effectiveness of alternative approaches to carbon storage. What scientists do know now is best summarized by land-use type.

Rural Forests

A key factor in the U.S. carbon cycle is the forest, a major land use in this country. *(See text Figure 3.)* In 1620, forests covered an estimated 45 percent or more of the land area of today's United States—more than 1 billion acres.[1] Soon thereafter, the wholesale clearing of forests for agriculture and pasture began.[2] Between 1700 and 1980, an estimated 1.7 million acres per year of forest were converted to agricultural land in North America.[3] In the United States alone, estimates of forest losses from 1600 to 1980 range from 600,000 to 1 million acres per year.

By the early 1900s, U.S. forest cover had dipped to a low point of about 750 million acres.[4] By the 1950s, the reversion of abandoned crop and pasture lands to forest cover, together with planned reforestation, returned total forest cover to more than 780 million acres. Since then, however, forest cover has declined to just 730 million acres, or 32 percent of the surface area of the United States.[5] Between 1977 and 1987, an average of almost 400,000 acres of forest was lost annually.[6] These losses probably resulted in the release of some 20 million tons of carbon during each of those years.

Statistics reflecting changes in timberland acreage can say more about changes in classification than about changes in carbon stocks.

Of course, forest area tells only part of the story of carbon stocks. Often, statistics reflecting changes in timberland acreage say more about changes in use than they do about changes in carbon stocks. For example, when land classified as timberland is reclassified as wilderness, the resulting decrease in timberland acres is not matched by a decrease in forested acres or carbon.[7] In other cases, timberland is redefined as ''other'' forest land when the site's productivity is reevaluated even though the number and type of trees might not change at all. Alternatively, a real change in the resource might not show up in forest-acreage statistics. If one million acres of mature forest were converted to agricultural land and one million acres of agricultural land elsewhere were converted back to forest or tree-crop cover during the same year, no change would appear in forest area statistics; indeed, net growth estimates might even rise soon after the switch. Yet it would probably take decades, if not centuries, for the carbon accumulating on the newly forested acres to equal the quantity of carbon lost when the mature forest was cut. *(See Box A-1.)* The same principle applies to the replacement of virgin forest with managed forests, a consistent trend in the United States since Europeans settled the continent. Indeed, commercial timber volumes in U.S. forests have declined by an estimated two-thirds since 1630, while forest area has declined by less than one-third. *(See Figure A–1.)*

The first estimates of total U.S. forest ecosystem carbon stocks suggest that some 57 billion tons of carbon are currently tied up in those ecosystems.

The first estimates of total U.S. forest ecosystem carbon stocks suggest that some 57 billion tons of carbon are currently tied up in those ecosystems.[8] The carbon loading of an acre of forest can vary widely, depending on species composition, age, geography, and soil type. On average, a forested acre in the United States is estimated to hold almost 80 tons of carbon in biomass and soils, a large proportion of which is lost when the land is cleared for alternative uses. Surprisingly, most of the

---

Box A–1. Changing Land Uses and Carbon Flows in the Southeastern United States

Few attempts have been made to track the impact of long-term land-use changes on carbon flows in a specific region. Delcourt and Harris, however, have done just that for the southeastern United States from 1750 to the present. They estimate that between 1750 and 1960 the total quantity of carbon in the soils and vegetation of the southeastern United States declined linearly from 47.6 to 16.6 billion tons as a result of converting forest and native prairie to agricultural uses. This change suggests an average annual net release of some 143 million tons of carbon per year, making the southeastern United States a major source of biotic carbon for most of the period since the industrial revolution.

Delcourt and Harris estimate that since 1960 the same region has become a net sink

for carbon as abandoned lands have reverted to forest and as forested lands have become more densely stocked and managed. They estimate, however, that only 77 million tons of carbon are being taken in on net every year, one-half as much as was lost for two centuries. Since 1960, therefore, only 3.1 percent of the carbon lost from between 1750 and 1960 has been restored. They conclude that because of the much higher biomass densities found in the original virgin forests, managed reforestation in the region will never fully offset losses incurred in previous years.

Source: Delcourt, H.R., and Harris, W.F., 1980. ''Carbon Budget of the Southeastern U.S. Biota: Analysis of Historical Change in Trend from Source to Sink,'' *Science* 210:321–23.

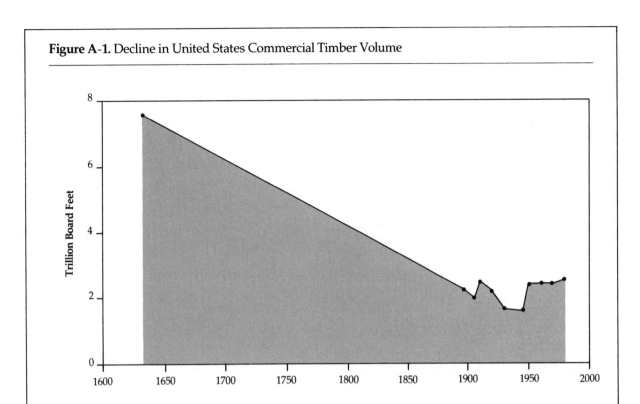

Figure A-1. Decline in United States Commercial Timber Volume

Trillion Board Feet

| | | | | | | | | |
| 1600 | 1650 | 1700 | 1750 | 1800 | 1850 | 1900 | 1950 | 2000 |

Source: M. Clawson, "Forests in the Long Sweep of American History," *Science*, 204, (1979): 1168–74

carbon in U.S. forest ecosystems appears to reside in forest soils, not in the trees themselves. *(See text Figure 6.)* Considering these findings and that nearly all of today's forest cover has been disturbed over time and thus is likely to contain significantly less carbon than it once did, a plausible estimate of the amount of carbon present in the biota at the time of European settlement is well over 100 billion tons. Indeed, according to Richard Houghton of the Woods Hole Research Institute, North American land conversions released an estimated 22 billion tons of carbon between 1860 and 1980 alone.[9]

Timberlands

As with forest cover generally, timberland area declined from the time of European settlement to 1920, bottoming out at approximately 468 million acres. Timberlands are commonly defined as "unreserved" forests capable of

producing more than 20 cubic feet of merchantable timber per acre per year, or about 0.4 tons of carbon. By 1962, timberland area had rebounded to 516 million acres as cotton lands in the South and hilly clearings in other areas reverted to forest. But between 1962 and 1977, timberland area fell five percent to 491 million acres, and from 1977 to 1987, timberlands dropped further to some 483 million acres (about 66 percent of total U.S. forested land), where they remain today. Of these, approximately 278 million are controlled by small-scale landowners, and less than 70 million acres are in the hands of industrial timber interests. *(See Figure A–2.)*

Various factors have contributed to declining timberland area, including conversion to agricultural and developed uses. Since 1962, for example, some 6.1 million acres of timberland have been converted to urban and agricultural uses in the Southeast.[10] An additional 35

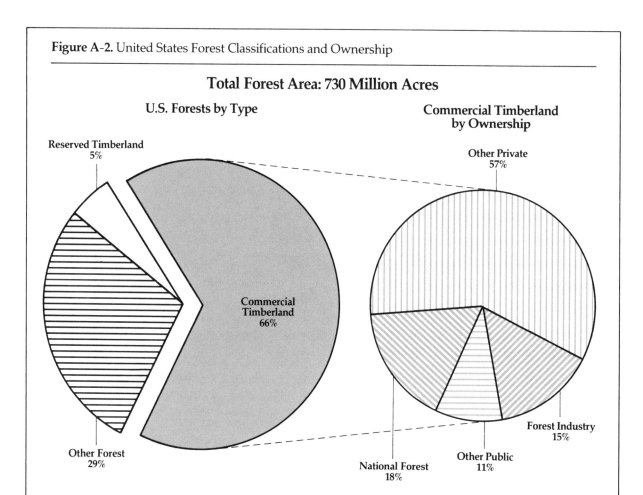

Figure A-2. United States Forest Classifications and Ownership

Total Forest Area: 730 Million Acres

U.S. Forests by Type

Reserved Timberland
5%

Commercial
Timberland
66%

Other Forest
29%

**Commercial Timberland
by Ownership**

Other Private
57%

Forest Industry
15%

Other Public
11%

National Forest
18%

Source: Based on U.S. Department of Agriculture, Forest Service, *An Analysis of the Timber Situation in the United States: 1989–2040, Part I: The Current Resource and Use Situation*, draft, (1988); and K.L. Waddel, *et al., Forest Statistics of the United States, 1987*, U.S. Department of Agriculture, Forest Service, Pacific Northwest Research Station, (1989)

million acres of forest that might meet the productivity requirement of timberlands are now included in reserves, including national parks and wilderness areas.

The United States Forest Service (USFS) projects that national timberland area will fall by an additional 20 million acres by 2040,[11] with most of this decline reflecting real changes in the resource rather than timberland withdrawals. Tree planting, however, has been on an upward trend for the last 50 years,

mainly because rates of artificial regeneration on harvested timberlands have been increasing. *(See Figure A-3.)* Some government programs are encouraging the afforestation of croplands too. *(See text Box 7.)* Reforestation programs tend to come and go, however, and land conversion under the Conservation Reserve Program (CRP) is scheduled to end soon as well. Although such programs could reverse the projected decline in timberland area, their long-term effect cannot yet be ascertained. *(See text Figure 4.)*

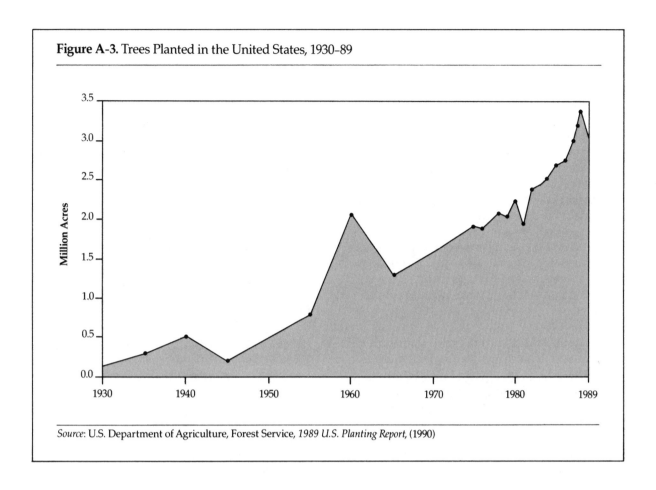

Figure A-3. Trees Planted in the United States, 1930–89

Source: U.S. Department of Agriculture, Forest Service, *1989 U.S. Planting Report*, (1990)

Despite declining timberland area, timber volumes on the remaining lands are increasing, according to USFS statistics. U.S. timberlands contained an estimated 830 billion cubic feet or more of merchantable timber in 1987—up more than four percent from the previous decade.[12] Most of these productivity gains have been realized on private nonindustrial timberlands, particularly the 164 million acres located in the southeastern United States. Overall, commercial timber growth rates exceeded rates of harvest by an estimated 20 percent in 1989 (by 34 percent on national forests).[13] As a result, some argue, U.S. timberlands can support significantly larger harvests, particularly on national forest lands.

Is it wise to increase harvest levels? Even on private timberlands, experts question the long-term sustainability of existing growth rates and rotation times.[14] Foresters are already concerned over unexplained decreases in growth rates on some timberlands. *(See Box A-2.)* As for the national forests, many analysts have argued that even logging at today's levels has often proven uneconomic, unsustainable over the long term, and too high in light of competing social priorities, including endangered species conservation and watershed protection. The Wilderness Society estimates, for example, that if a long-term sustainable yield criterion were integrated into multiple-use objectives, annual sales from the national forests would not exceed 5.0 to 6.5 billion board feet of timber, considerably less than the 10.5 billion board feet that the Forest Service originally expected to sell in fiscal year 1991.[15]

Other Rural Forests

Of the 245 million acres of the United States now classified as unreserved forests but not as timberlands, almost 60 percent are accounted for by the fir-spruce forests of interior Alaska and the pinyon-juniper forests of the Rockies.[16] Relatively little is known about how these economically less important forests have changed over time, much less about trends in their carbon flows. Their potential role in U.S. biotic carbon balances is therefore hard to assess.

Crop, Pasture, and Rangelands

Agricultural, pasture, and rangelands account for just over one billion acres, or 47 percent of

the land area of the United States.[17] This figure, down by approximately 50 million acres since 1949, is falling gradually.[18] Significant quantities of agricultural land continue to be converted to urban and other developed uses, and many of the acres currently dedicated to agriculture and range were once far more richly laden with biomass than they are today. An estimated 250 million acres of forest land have been converted to agricultural and pasture use since North America was settled. In the process, most standing biomass on those acres was eliminated, resulting in the release of enormous quantities of carbon to the atmosphere.

Conversion of land from one use to another continues to this day in the United States, though not on the scale seen in past years. More than one-half of the 12 million acres of wetlands drained for agricultural uses from the mid-1950s to the 1970s were forested.[19] When such lands are stripped of tree cover to make way for agriculture, loss of standing biomass is not the only source of carbon released into the atmosphere. Soil carbon levels fall by an estimated 40 percent or more.[20] Erosion compounds this problem: an estimated 44 percent of all U.S. cropland, or some 170 million acres, is eroding at rates sufficient to destroy its long-term productivity, and fully one-quarter is losing more than 19 tons of topsoil per acre per year.[21] In 1982, total soil losses from U.S. croplands exceeded 3 billion tons per year. Total soil losses from all nonfederal rural lands, primarily through water and wind erosion, approached 5.5 billion tons in the same year. Although not all this carbon will oxidize to CO_2, some eventually will. In addition, the depletion of soil productivity owing to topsoil loss often requires an increase in the application of nitrogen fertilizers, which require great amounts of energy to manufacture—a process that emits CO_2—and which can trigger emissions of other greenhouse gases, such as nitrous oxide (N_2O).

Urban Forests and Developed Lands

Historically, the need for agricultural land has driven most conversion of forest land. Recently, however, land conversion to urban and other developed uses such as highways has become more and more important. Unfortunately, though, discrepancies in how different government agencies measure and define urban areas make it difficult to arrive at a clear picture of the situation.[22] The Department of Agriculture estimates that the average annual conversion of rural land to developed uses increased from 0.9 million acres per year in the 1960s to 1.3 million acres per year by 1980 and that land committed to urban areas, highways, airports, and other such uses now totals over two percent of the U.S. land area.[23] (See Figure A–4.)

The expansion of developed land uses is usually at the expense of woodlands or croplands. About one-half million acres of each are lost to development annually.[24] In the Southeast alone, 170,000 acres of timberland are converted to urban use each year.[25] In Fulton County outside Atlanta, forest cover has decreased by 50,000 acres in the past 15 years; in 1988 alone, an estimated 50 acres of forest and cropland were lost there every day. Similar losses are occurring outside Washington, D.C., where 80,000 acres of forest have been eliminated in Fairfax County, Virginia, in the past 35 years.[26] As we move into the 1990s, the USFS projects, forest conversion will stem primarily from expanding needs for reservoirs, urban areas, highways, airports, and mining.[27]

As we move into the 1990s, forest conversion will stem primarily from expanding needs for reservoirs, urban areas, highways, airports, and mining.

The loss of forested land, whether directly or indirectly the result of urban development, reduces U.S. carbon reservoirs. Significant quantities of biomass and soil carbon are lost each year, even when some trees remain or are replanted in cities. In fact, the American Forestry Association estimates that 61 million trees line urban streets and another 550 million trees grow elsewhere in urban or builtup areas.[28] Comprehensive surveys of our urban forests are not available, but it is known that a large fraction of these trees are relatively small (less than 12 inches in diameter) and that the habitat for urban trees often is poorly planned and maintained. Urban trees are also under assault from various airborne pollutants. A common rule of thumb is that four urban trees are dying for every new tree planted around the nation.

The decline of our urban forest may soon be reversed, at least temporarily. Billed as something

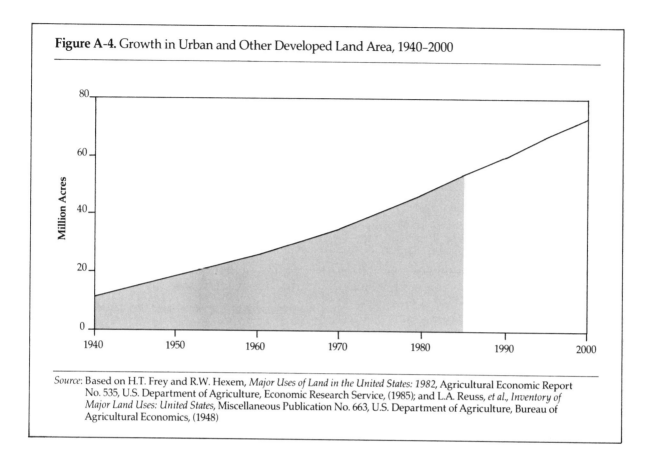

Figure A-4. Growth in Urban and Other Developed Land Area, 1940–2000

Source: Based on H.T. Frey and R.W. Hexem, *Major Uses of Land in the United States: 1982*, Agricultural Economic Report No. 535, U.S. Department of Agriculture, Economic Research Service, (1985); and L.A. Reuss, *et al.*, *Inventory of Major Land Uses: United States*, Miscellaneous Publication No. 663, U.S. Department of Agriculture, Bureau of Agricultural Economics, (1948)

individuals can do to combat global warming while making the local environment more comfortable and attractive, urban tree planting has attracted a large following recently. Global Releaf programs have been initiated by the American Forestry Association in 11 states and more than 100 communities. Other programs and organizations—among them, Trees Atlanta, Tree People (Los Angeles), and Philadelphia Green—are seeking similar ends, and the president's America the Beautiful program is likely to spur federal involvement in urban forestry. (*See text Box 6.*)

Consumption of Wood Products

An estimated 42 billion cubic feet of wood, the equivalent of 390 million tons of carbon, are removed from U.S. forests annually.[29] Of this quantity, only a fraction is used for durable wood products.[30] Production of roundwood products from U.S. forests was estimated at 18 billion cubic feet for 1986,[31] the equivalent of some 165 million tons of carbon. The disposition of all this carbon is a potentially important element of the biotic carbon cycle because wood products can serve as an artificial carbon sink by keeping the harvested carbon out of the atmosphere for months (in the case of disposable products) to many decades (in the case of construction materials and other long-lived items).

Although the average useful life of wood products can be reasonably well estimated, the carbon storage of these products remains difficult to ascertain. It is known that a huge stock of worn-out wood products can now be found in U.S. landfills and that decomposition there is slow, but research on how much carbon is released in the utilization and disposal of wood products and the mechanisms through which it is released is just beginning.[32] It remains unclear, therefore, what the size of this artificial carbon reservoir is and whether it is now a net sink or source of carbon. As the quantity of stored carbon builds up, for example, even gradual decomposition will eventually release more carbon than can be stored in the form of new wood products. In landfills, decomposition can also release methane, a much more potent greenhouse gas than carbon dioxide.

Wood products are by no means the only product of U.S. forests that figure into global warming. By at least one estimate, more than one-half the wood removed from U.S. forests in the early 1980s was burned for energy,[33] and fuelwood offtake from forest lands is expected to increase significantly in coming years.[34] Residential fuelwood consumption, for example, is expected to increase by 64 percent between 1986 and 2020.[35] Right now, direct combustion of wood and wood wastes currently accounts for almost 90 percent of all biomass energy used in the United States but only 4 percent of total energy consumption and 8 percent of industry's energy use.[36] Most of this energy is consumed in the paper-and-pulp industry, where it supplies 70 percent of all energy requirements. This use is likely to grow three- to fourfold over the next several decades.[37]

Conclusions

Although land uses in the United States are relatively well tracked, major uncertainties remain with respect to the net carbon flows associated with the changes that continue to take place. It is virtually certain that forests, especially timberlands, are the nation's largest storehouses of biotic carbon, but a variety of questions need answers before the size and direction of net carbon flows can be estimated:

- How large are stocks of noncommercial biomass on timberlands? And how much carbon is stored in forested lands not classified as timberlands? What are the trends in each area?

- Are net changes occurring in the stresses facing existing forests and timberlands? Are rising fire frequencies resulting in net releases of forest carbon? Will air pollution continue to slow the growth of large areas of U.S. forest?

- How much carbon is stored in urban forests? How are trends in urban forestry affecting fossil fuel consumption?

- How much carbon is stored in wood products (both in use and disposed of)? Have we already exceeded the point at which annual oxidation of a small proportion of these products equals annual additions to product stocks?

- If the billions of tons of carbon previously thought to be disappearing into the world's oceans are instead fertilizing temperate forests around the world, what are the impacts on domestic forests?

These and other uncertainties notwithstanding, one key fact is incontrovertible. As hundreds of millions of acres of forests in what is now the United States were cut down over the last 200 years and converted to agriculture, pasture, and urban uses and as virgin forests were replaced with younger forests and plantations, tens of billions of tons of carbon were added to the atmosphere.

Notes

1. Forest area in the United States has been specifically tracked since 1952, and there is general agreement on the current expanse of U.S. forests. Estimates of forest cover earlier in the century and in prior centuries in what is now the United States, however, vary significantly. Sedjo, for example, estimates original forest cover at 950 million acres, but the Forest Service's estimate is 1.1 billion acres. Sedjo, R.A., 1990. *The Nation's Forest Resources.* Discussion Paper No. ENR90–07. Resources for the Future; U.S. Department of Agriculture, Forest Service, 1989. *Tree Planting and Forest Improvement to Reduce Global Warming.*

2. For a discussion of the history of U.S. forests, see Williams, M., 1988. ''Death and Rebirth of the American Forest: Clearing and Reversion in the United States 1900–1980,'' in Richards, J.F., and Tucker, R.P., eds. *World Deforestation in the Twentieth Century.* Duke University Press.

3. Houghton, R.A., et al., 1983. ''Changes in the Carbon Content of Terrestrial Biota and Soils Between 1860 and 1980: A Net Release of CO_2 to the Atmosphere,'' *Ecological Monographs* 53:235–62.

4. U.S. Department of Agriculture, Forest Service, 1989.

5. U.S. Department of Agriculture, Forest Service, 1990a. *America the Beautiful: National Tree Planting Initiative.*

6. Available statistics invite different interpretations of trends in forest acreage. Some USFS data, for example, suggest that the United States lost, on net, 700,000 acres of forested land per year during the last decade. This is the figure used in the president's 1991 budget proposal. Most foresters, however, consider the figure too high, preferring an average rate of loss of approximately 400,000 acres per year. Birdsey, R.A., pers. comm.

7. Between 1977 and 1987, for example, eight million acres of timberland were apparently lost in the United States. Further analysis, however, shows that much of the recorded loss of timberland was due to redefinition of land use (i.e., from national forest to national wilderness area) rather than an actual reduction in forested lands. In Alaska alone, nearly four million acres were placed in reserved status as national parks and wilderness from 1977 to 1987.

8. The estimate is based on the extrapolation from timber inventory statistics to total tree biomass, the modeling of forest soil carbon on the basis of temperature, precipitation, and forest-stand age, and empirical data on biomass in understory vegetation and litter for selected ecosystems. Nevertheless, it represents only the first attempt to quantify these variables and may be subject to significant change as research progresses. Birdsey, R.A., in press. ''Potential Changes in Carbon Storage Through Conversion of Lands

to Plantation Forests," in *Proceedings, North American Conference on Forestry Responses to Climate Change,* 1990, Washington, D.C. Climate Institute.

9. Houghton, R.A., et al., 1983.

10. U.S. Department of Agriculture, Forest Service, 1988a. *The South's Fourth Forest: Alternatives for the Future.* Forest Resources Report No. 24.

11. U.S. Department of Agriculture, Forest Service, 1990b. *An Analysis of the Timber Situation in the United States: 1989–2040.* Editorial draft.

12. Because timber statistics focus almost exclusively on merchantable timber, it remains unclear to what degree these increases reflect actual increases in biomass as opposed to a reallocation of biomass among commercial and noncommercial species.

13. Mickey, R.S., 1990. Statement before the House Agricultural Committee, Forests, Family Farms, and Energy Subcommittee. February 6.

14. Robinson, G., 1988. *The Forest and the Trees: A Guide to Excellent Forestry.* Island Press.

15. Wilderness Society, 1989. *New Directions for the Forest Service: Instructions from the President of the United States to the Chief of the Forest Service (Recommendations of the Wilderness Society).*

16. U.S. Department of Agriculture, Forest Service, 1988b. *An Analysis of the Timber Situation in the United States: 1989–2040. Part I: The Current Resource and Use Situation.* Draft.

17. All land-use estimates presented here are approximate. Although several government agencies are charged with periodically reviewing nationwide land uses, differing organizational goals as well as inconsistencies in definition and measurement methods often result in land-use estimates that are difficult to compare. Even within the same agency, estimates made from one decade to the next are often not easily comparable.

18. Frey, H.T., and Hexem, R.W., 1985. *Major Uses of Land in the United States: 1982.* Agricultural Economic Report No. 535. U.S. Department of Agriculture, Economic Research Service, Natural Resource Economics Division.

19. U.S. Department of Agriculture, 1986. *Environmental Assessment for the Regulations Implementing the Wetland Conservation Provisions of the Food Security Act of 1985.*

20. Delcourt, H.R., and Harris, W.F., 1980. "Carbon Budget of the Southeastern U.S. Biota: Analysis of Historical Change in Trend from Source to Sink," *Science* 210:321–23; Covington, W., 1981. "Changes in Forest Floor Organic Matter and Nutrient Content Following Clearcutting in Northern Hardwoods," *Ecology* 62(1):41–49.

21. U.S. Department of Agriculture, 1989. *Conservation Reserve Program: Progress Report and Preliminary Evaluation of the First Two Years.*

22. Even within the same agency, changing methodologies can add to the confusion. The 1977 Soil Conservation Service (SCS) National Resources Inventory, for instance, used two sampling methods to estimate urban land use. The first, which used county maps to measure urban and builtup areas larger than 40 acres, was later determined to have overestimated builtup areas by several million acres. To correct this sampling error, the 1982 National Resources Inventory (NRI) changed measuring procedures and calculated urban area entirely from sample data. As a result, the NRI statistics now show acreage devoted to urban and transportation uses increasing from 51 million acres in 1958 to 90 million acres in 1977 and then falling to 73 million acres in 1982.

23. Conservation Foundation, 1987. *State of the Environment: A View Toward the Nineties.*

24. Bowen, G., 1989. Statement before the House Committee on Agriculture, Subcommittee on Forests, Family Farms, and Energy. June 7.

25. U.S. Department of Agriculture, Forest Service, 1988a.

26. Moll, G., 1989. ''The State of Our Urban Forest,'' *American Forests*, November/December:61–64.

27. U.S. Department of Agriculture, Forest Service, 1988c. *An Analysis of the Timber Situation in the United States: 1989–2040. Part II: The Future Resource Situation.* Draft.

28. Moll, G., 1989.

29. Birdsey, R.A., in press.

30. For example, Harmon et al. estimate that 42 percent of timber currently harvested in the Pacific Northwest is converted to long-term products, compared with 20 percent in the 1950s. An even smaller proportion of smaller trees would be likely to end up in long-lived products. Harmon, M.E., Ferrell, W.K., and Franklin, J.F., 1990. ''Effects in Carbon Storage of Conversion of Old-Growth Forests to Young Forests,'' *Science* 247:699–702.

31. U.S. Department of Agriculture, Forest Service, 1988b.

32. Row, C., and Phelps, R.B., in press. ''Carbon Cycle Impacts of Improving Forest Products Utilization and Recycling,'' in *Proceedings, North American Conference on Forestry Response to Climate Change*, 1990, Washington, D.C. Climate Institute.

33. U.S. Congress, Office of Technology Assessment, 1983. *Wood Use, U.S. Competitiveness and Technology.* OTE–E–210 U.S. Government Printing Office.

34. U.S. Department of Agriculture, Forest Service, 1990b.

35. High, C., and Skog, K., 1989. ''Current and Projected Wood Energy Consumption in the United States.'' Paper Presented at the Energy from Biomass and Wastes XIII Conference, New Orleans, 1989.

36. U.S. Department of Energy, Energy Information Administration, 1989. *Annual Energy Review.* DOE/EIA–0384(89); Rader, N., 1989. *Power Surge: The Status and Near-Term Potential of Renewable Energy Technologies.* Public Citizen.

37. High, C., and Skog, K., 1989.

World Resources Institute

The World Resources Institute (WRI) is a policy research center created in late 1982 to help governments, international organizations, and private business address a fundamental question: How can societies meet basic human needs and nurture economic growth without undermining the natural resources and environmental integrity on which life, economic vitality, and international security depend?

Two dominant concerns influence WRI's choice of projects and other activities:

The destructive effects of poor resource management on economic development and the alleviation of poverty in developing countries; and

The new generation of globally important environmental and resource problems that threaten the economic and environmental interests of the United States and other industrial countries and that have not been addressed with authority in their laws.

The Institute's current areas of policy research include tropical forests, biological diversity, sustainable agriculture, energy, climate change, atmospheric pollution, economic incentives for sustainable development, and resource and environmental information.

WRI's research is aimed at providing accurate information about global resources and population, identifying emerging issues, and developing politically and economically workable proposals.

In developing countries, WRI provides field services and technical program support for governments and non-governmental organizations trying to manage natural resources sustainably.

WRI's work is carried out by an interdisciplinary staff of scientists and experts augmented by a network of formal advisors, collaborators, and cooperating institutions in 50 countries.

WRI is funded by private foundations, United Nations and governmental agencies, corporations, and concerned individuals.